Praise for *Project Animal Farm*

"Written in the most vivid and engaging way, this remarkable book demands to be read by anyone who cares about where their food comes from."

—Philip Lymbery, CEO of Compassion in World Farming and author of *Farmageddon*

"An extraordinary triumph of narrative nonfiction . . . a seminal work of investigative journalism."

—David Kirby, author of *Animal Factory*

"A poignant work of depth and clarity. An accessible and educational examination of our food system. Highly recommended."

—Gene Baur, President of Farm Sanctuary

"A work of unrelenting passion, *Project Animal Farm* is a very compelling and informative read."

—Roger Harley, Farmer

"*Project Animal Farm* is amazing. Written with sparkling narrative and infectious passion."

—Kabir Sehgal, author of *Coined*

"An eyes-wide-open view of industrial animal farming and its effect on farm animals and on consumers. A compelling and thoughtful work, a clarion call for acknowledgment."

—Wayne Pacelle, President, The Humane Society of the United States

"Filled with deeply personal stories and facts about food production, *Project Animal Farm* is a game-changer."

—Marc Bekoff, ethologist and author of *The Emotional Lives of Animals*

"An engaging expedition that exposes both the animal suffering and the urgent public health threats posed by factory farms."

—Aysha Akhtar, author of *Animals and Public Health*

"Sonia Faruqi shines a brave light on what, by many estimates, is the most pressing issue of our time. It is required reading for anyone who eats food."

—Jonathan Balcombe, author of *Pleasurable Kingdom*

"Warning: This book might change your life!"

—Hal Herzog, author of *Some We Love, Some We Hate, Some We Eat*

"Compassionate, engaging and wise, Sonia Faruqi guides readers through a polarizing subject—the lives of farm animals—and takes us to places that challenge our ideas, shed light on complexities, and propel new thinking."

—Lorraine Johnson, author of *City Farmer*

"Compelling, fascinating, empowering and hopeful, *Project Animal Farm* will forever change the way you look at farm animals and will help you become a part of the dramatic and long overdue change the book demands."

—Rob Laidlaw, biologist and children's book author

"A perfect combination of excellent writing and in-depth understanding."

—Wayne Roberts, author of *The No-Nonsense Guide to World Food*

"Faruqi's bold, critical, and ultimately hopeful quest for ideal animal agriculture reminds us that how we treat our fellow beings is a deeply moral matter."

—Tovar Cerulli, author of *The Mindful Carnivore*

"An engaging, penetrating tour. Insightful."

—Katherine Gustafson, author of *Change Comes to Dinner*

"Every so often a book comes along that has the power to alter the course of history. *Project Animal Farm* is that potent. It is so inspiring, so moving, so deeply personal and yet also has such profound cultural implications, that it will change the lives of everyone who reads it. People will be talking about this book for decades."

—John Robbins, author of *The Food Revolution*

"Into an engaging account of the adventures of a young city dweller among the factory farmers, Sonia Faruqi manages to smuggle a body of useful—and disturbing—information about this most secretive of global enterprises."

—J. M. Coetzee, Winner of the Nobel Prize in Literature

"Everybody who is interested in food policy and animal welfare should read this book. *Project Animal Farm* will make you think long and hard about problems with intensive dairy farms and egg production."

—Temple Grandin, scientist and author of *Animals in Translation*

"Let Sonia Faruqi take you on her courageous personal journey of discovery. I thought I already knew everything there is to know about modern animal production, but I learned many new things from this very readable book, and you will, too."

—Peter Singer, philosopher and author of *Animal Liberation*

"Brave, captivating, enlightening, and impossible to put down, this remarkable true story pries open hearts and minds and exposes the travesty of industrial farming like none other."

—Frances Moore Lappé, author of *Diet for a Small Planet*

"This is a book about the people who make their living putting food on our tables. In it, you'll see how some of them struggle to do it well, while others are willing to do it any way at all if it will help them get ahead. You'll be glad to have seen them all with Sonia Faruqi's eyes."

—Daniel Quinn, author of *Ishmael*

PROJECT ANIMAL FARM

AN ACCIDENTAL JOURNEY INTO THE SECRET WORLD OF FARMING AND THE TRUTH ABOUT OUR FOOD

SONIA FARUQI

PEGASUS BOOKS
NEW YORK LONDON

PROJECT ANIMAL FARM

Pegasus Books LLC
80 Broad Street, 5th Floor
New York, NY 10004

First Pegasus Books cloth edition July 2015

Interior design by Maria Fernandez

Library of Congress Cataloging-in-Publication Data is available.

ISBN: 978-1-60598-798-9

10 9 8 7 6 5 4 3 2 1

Printed in the United States of America
Distributed by W. W. Norton & Company

To my twin sister, Sofia,
for all her love and support

CONTENTS

PROJECT ANIMAL FARM

PROLOGUE

THE START OF A JOURNEY

A MATERIALISTIC INVESTMENT BANKER

Growing up, I seemed almost earmarked for the financial world, as I was both studious and materialistic. I was a small mass of seriousness—a little librarian—completing my assignments early and happily. I preferred class to recess, homework to hopscotch. In middle school, I would for hours paint still lifes—a bowl of apples or a vase of flowers—feeling that my own life was a still life—a cantaloupe compressed into a teacup. My core longed for more, mentally, materially.

My wallet was wide enough to fill the well of my needs—food, school, shoes—but not the valley of my whims and wishes. I craved more clothes, more books, more boxes—dainty little tins in which to cherish my dainty little things. My mother occasionally said, "You like having things for the sake of having them; you have the joy of possession." I interpreted her comments as compliments: how sophisticated I was, I thought, to be possessed with the joy of possession.

After high school, I attended Dartmouth College, a small, liberal arts university with a campus dotted with trees. Each of its classes was an ingredient you were supposed to select, slice, and stir in the pot of your mind, toward the aim of not merely faring but flourishing. Dartmouth urged that for "bonding" purposes, its students hike together in small groups for five days before the start of school. I'd never hiked before—because I'd never wanted to, and I'd grown up in city apartments—but I decided to participate because everybody else did. Two hours into my outdoor excursion, however, I realized that it was, as I termed it then, "the biggest mistake I've ever made."

My backpack felt like a rock strapped to my back, despite the fact that I'd reluctantly unloaded my makeup and chocolates at the insistence of trip leaders. Worse, there was nothing for me to eat: I ate only meat—shunning fruits and vegetables since childhood—and there was no meat, as it would have spoiled over the course of the trip. I decided to stop eating. An even more pressing issue than food was facilities. There were no toilets; we were supposed to go in the woods like chest-pounding cavemen. I decided to hold it for days. (And I did.)

Finally, there was the wildlife: I felt sure a bear would attack me as I slept. One night, I thought I heard an animal panting and salivating right beside my ear. "GIVE ME YOUR FLASHLIGHT!" I yelled at the snoring, sleeping bag–encased form next to me, rattling him awake. I flashed his light everywhere. But there was no bear; there were only my trip members, awake and annoyed. I resolutely avoided the outdoors after that. The rugged life clearly did not suit me.

I graduated from Dartmouth College with a major in economics and public policy, and a minor in government. Degree in hand, I joined my classmates in a stampede to Wall Street. Wall Street was the money business, the fast track, a meal ready to eat, without the costly condiment of a graduate degree. I felt pleased and fulfilled when an investment bank offered me employment.

Investment banks had just one requirement of their young employees. They required that, as the moon revolves around earth, employees revolve around work. As an investment banker,

I did not work to live, I lived to work. I did not eat to live, I ate to work—and I ate at work. All of my meals—breakfast, lunch, and dinner—I consumed in my cubicle, gobbling them up rapidly so that I could continue typing, calculating, working: a machine in the form of a woman.

But I enjoyed it. I liked feeling important. I liked having a paycheck. I liked wearing a suit every day. I liked strutting across the office carpet in high heels, papers under my arm. I liked my Upper East Side apartment, only a short walk from Central Park. I liked racing my fingers across my keyboard and my eyes across my computer screen. Every morning, I awoke like a golden retriever puppy salivating to start the day, unfazed by my seventy-hour work weeks. Wall Street was where I was meant to be, I felt.

Until I was let go.

The American economy began hemorrhaging immediately after I joined, starting with the subprime mortgage sector. The pain spread outward, until the entire financial system convulsed in its throes. Investment banks decreed that it was no longer profitable for them to continue to feed the small fry they'd lured into their nets with baits of bonuses. So they cut the nets. They laid off hundreds of thousands of employees. After two years of living in my cubicle, I was forced to leave it.

I planned to apply for roles at other financial firms. I would continue to burn with the same fire, only its cinders would be raked by another bank. My life would remain the same gift basket of enjoyment, tied with the green ribbon of Excel, adorned with the red bow of PowerPoint, but would shift to a new cubicle. *After* a break, though. I'd worked nonstop on Wall Street, taking not even one sick day, and I thought a break would recharge my battery and help me recommence with renewed intensity.

During my break, I read books. I contemplated life. And I moved from New York to Toronto.

Since I had plenty of time on my hands, I decided to volunteer at a farm, imagining that the experience would be an adventure. I contacted a dozen organic and small farms with an enthusiastic offer of free assistance toward the production of their food. I was sure they'd be thrilled and grateful. They weren't. Most of them were cold and uninterested. Only one—an organic dairy farm—accepted my ambiguous offer, and only under one condition: that I volunteer with them not for a week at the most, as I'd hoped, but for two weeks at the least.

I reluctantly agreed to the duration. The dairy farm reluctantly agreed to supply accommodation.

I'd devoured the *Little House on the Prairie* books as a girl, and my mental image of organic farms resembled the pastoral, prairie-like setting of the books. I imagined that my farm stay would be both an education and a vacation, and that, shortly after, I would return to the prosperous world of suits, spreadsheets, and skyscrapers.

I had no idea what I was getting myself into.

PART I

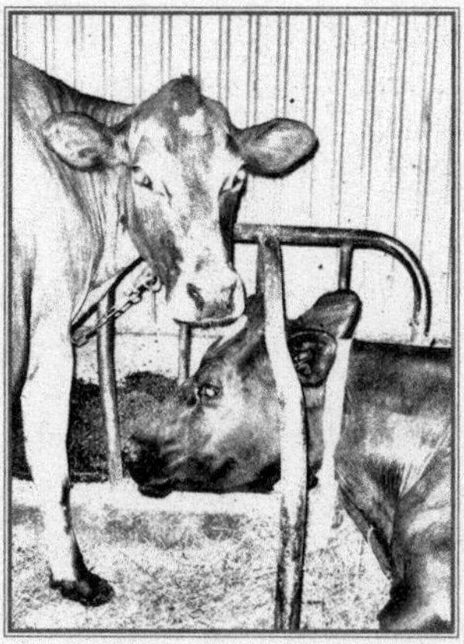

CHAPTER ONE

ORGANIC DAIRY COWS

TETHERED, TRAINED BUTTERCUPS

Michael and Irene Miller were baffled to learn that I did not know how to drive a car, let alone a truck or tractor.

"You're how old?" Irene asked me.

"Twenty-five."

"So how come you can't drive?"

"I take the subway."

Michael and Irene looked dissatisfied with my response. They were also confused to learn that I did not know anything about farms—not even the difference between hay and straw. They claimed that this was like not knowing the difference between bread and bed. Hay was like bread—to eat—and straw was like a bed—to sleep. Would I ever sleep on a slab of bread or eat my bed?

I would not, I assured them, embarrassed.

I was further embarrassed by my appearance. Irene Miller wore not a spot of makeup or jewelry, whereas I was wearing dangling earrings, bright blush, and a colorful top. I looked dressed for a party.

Irene seemed desperately eager to say something. "Everyone around here is white," she told me in her Dutch accent, staring at my brown skin and black hair. "A few years ago, a black family moved in. Not to our village, but to a nearby one. We never saw them, but we heard a lot about them because everyone stared at them and followed them. No one liked them. They stayed for only a little time. Since then, no other black family—or Jewish or Chinese or Indian, like you—has come to our community. Everyone here is white."

I did not correct Irene that my background is Pakistani, not Indian. I lived in downtown Toronto only three hours away from the Millers, but I felt like I'd arrived in a foreign country.

For a time, the only sounds at the table were the scrapes of forks and knives against plates. Eventually, Michael informed me, "We have a hundred and thirty dairy cows and calves."

"I don't like animals," Irene rushed to contribute. She took a long, slow sip of wine directly from her wine bottle—her nightly routine, I would soon learn—and then glanced at the speckled, cinnamon Australian Shepherd to the other side of the room. "I didn't want to get a dog. Michael did, so we have one."

The dog had been hand-selected by Michael out of a litter of six puppies based on only one criterion: "He was the quietest one." The canine hush suited Michael because he was himself the quietest of his brood of brothers, growing only more silent over the years, becoming so taciturn that one had to wonder whether he'd taken an oath of silence. He was tall and gaunt, with an angular face lit by overcast, sky-blue eyes and etched with a cloud-white beard.

Like all of his ancestors in the Netherlands, Michael was a dairy farmer, though he'd once thought he might choose a different path. He'd studied tropical agriculture to "save the world." He never ended up working in the plantations of tropical nations, but he did make it out of the thirty-mile patch of the Netherlands in which his forefathers had passed the last four centuries, and in which Michael predicted that his four brothers and their descendants may pass the next four centuries, anchored there on that spot of land like mountains until eternity.

A few years into their marriage, Michael and Irene Miller* moved to Canada, to the two-story brick-and-wood house they continued to reside in today. They farmed the conventional way for their first ten years in the country, until Irene—whose hair had then, in winters, roared red like saffron, and in summers melted into the coral color of a setting sun—had an idea that radically altered their circumstances. She noticed that organic, a movement that started sputteringly after World War II as a counter-effort to agricultural industrialization, was climbing in sales, buoyed by consumer interest in health, sustainability, and animal welfare. Spotting a gap between organic demand and supply, she suggested to her husband that they get their dairy farm certified organic.

Certification was expensive, but Irene's decision was more lucrative than she'd ever imagined. In the twenty-plus years since the Millers had obtained organic certification, the organic sector had lifted off into the sky like a hot-air balloon, its sales reaching $35 billion in the United States in 2013. Today, organic products can be found in three quarters of American and Canadian grocery stores. Surveys show that almost half of Americans actively try to

* Names of individuals have been changed to protect personal privacy.

include organic foods in their diet, and two out of five Canadians buy organic products every week.

But Irene—whose hair was now the aged, flaxen yellow of scorched sand—was no longer glad about the success of her organic plan. She was sad, beset by pangs of regret and resentment.

As soon as he finished dinner, Michael stood up and, without a word, strode out the back door of the house. Irene, who was still chewing the (non-organic) lettuce salad and ham-and-bean chili she'd prepared for dinner, leaned closer to me. "It's my fault," she whispered, her face defeated. "It's all my fault. I wish I had had the idea for organic. I want to get out of dairy farming, but now we're stuck with the farm. It's harder to sell an organic farm than a regular farm."

"Why do you want to get out of dairy farming?" I asked, surprised and uncomfortable at the personal nature of the confession.

"Because I hate it. I am completely sick of it. We've had it for thirty years. Michael promised me he would sell the farm, or give it to our daughter Annie, before we became fifty-one. Then, when we became fifty-one, he promised me he would sell the farm before we became fifty-three. Then, when we became fifty-three, he said fifty-five. Then he said fifty-seven. But he will become fifty-seven next week, and he told me to give him two *more* years to sell the farm. But I can't wait two *more* years. I'm thinking of leaving without him. I hate this dairy farm."

Irene and I cleaned up after dinner without a further word; then I retired to my room upstairs. The room was rusty, dusty, and musty, furnished with an ancient television and stereo, two bookcases colonized by long-legged spiders, and a bed of worn, unwashed blankets. The season was winter, but the room was unheated, offering little insulation from the wind whistling outside. I'd been thinking it at dinner, but as I tossed and turned in bed, I felt certain that coming here was a mistake.

I'd rarely seen cows up close before, and so I visited the cowshed eagerly on my first morning with the Millers.

Sixty-five cows lived in the shed. Most of them were collages of black splashed with white. They were giants, weighing 1,300 pounds on average, and standing nearly six feet tall, their hooves, elevated at the back, creating the effect of high-heeled shoes. Their pink udders looked like bulbous balloons, veined with weight. Even their eyelashes were large, arching an inch around dark eyes.

As I entered, the cows stood up one by one, like saluting soldiers. When I extended a hand to them, some treated it like wilted hay, taking a whiff then turning away. A few perceived it as dewy grass, giving it a swift sniff and lick. Others regarded it as a cumbersome fly and shook it off. Many viewed it as an encroaching weapon. They leapt to their feet in a great flurry of flesh and attempted to vault back. But they couldn't: they were shackled to stalls by neck chains. I could still have touched the cows, or punched them, or poked their eyes, and their chains would have arrested them instead of me.

Each cow lived as cramped in her stall as a big foot in a small shoe. She spent her hours engaging in one of four activities. She would eat, her mouth rotating in a slow, circular motion as she chewed through the pile of corn and hay dumped daily before her. She would drink, from a soup-sized blue bowl that, when touched with her mouth, filled automatically with water from a labyrinth of overhead pipes. She would nuzzle—passing her head through from underneath the bar separating her from her neighbor—and she would lick her neighbor or else rest her head on her neighbor's neck. Finally, she would groom—rather, *try* to groom.

Though every cow's hindquarters were caked with a crusty layer of excrement, she was helpless to clean them. Her neck chain held her in place. In addition, just behind her back hooves lay a "manure gutter," a low channel that lined the rear of her stall, and into which she feared falling. If that weren't enough to quell all movement, directly above her shoulders dangled a device that Michael Miller called a "shit trainer."

A shit trainer is a zigzag-edged metal rod that punishes the cow underneath with a jolt of electricity whenever she does not position herself precisely at the stall-gutter boundary as she defecates. Trainers are painful and restrict movement severely, causing stress and nervousness in cows, for which reason they are banned in Sweden and parts of Germany.

Defecation was tedious for the Miller dairy cows. A cow would drag her back legs out of her muck in the manure gutter and would heave herself up until she was standing. She would then raise her tail and lift her shoulders, but *zing*!—her shoulders would brush the trainer. She would comply with the trainer's electric warning that, in her present position, her excrement would fall into her stall, and she would carefully step back a hoof-width. Her hind hooves now perching precariously at the very edge of her stall, she would defecate, fearing the whole time that she would lose her footing and fall.

None of this was what I'd expected of an organic farm. The Miller website was merry, with its colorful pictures of cartoon cows, but the cowshed was melancholy, with its two rows of feces-smudged cows in gray stalls facing gray walls. The neck tether, the cow trainer, and the manure gutter together trapped the cows, subduing them from ahead, from above, from behind. Cows could not even turn their heads around fully. They were numbered, ordered milk machines.

Danielle and Ken liked to call them "cattle beasts."

Blond, blue-eyed, eighteen-year-old Danielle was a minimum-wage worker at the Miller milk plant. She was a high school graduate, but a grudging one, for all she'd ever learned in classrooms was "School is *so* boring." In retrospect, Danielle wished she'd spent her formative years wielding her gun instead of her pen. "My funnest thing to do is take my gun and shoot pigeons," she told me cheerfully.

Ken was a good-looking, green-eyed, brown-haired, twenty-nine-year-old father of two boys. He regretted his high school years even more than Danielle did. "The piece of paper you get when you graduate doesn't help you make money or anything," he said. Money was also Ken's complaint against his current job, for the wage he now earned was less than the eighteen dollars per hour he'd earned as a night janitor at his previous place of employment, Campbell Soup Company.

As a volunteer at the Miller farm, I assisted Ken and Danielle whenever they requested, attired, like them, in a black shirt, hairnet, and green gloves. Guided by their congenial tutelage, I stamped containers with expiration dates, placed them in cardboard boxes, and taped and stacked the boxes. I'd never done such repetitive, mindless work before, and I much preferred spending time in the cowshed. So did Annie Miller, who became my favorite person at the farm.

Thirty-one-year-old Annie was the only one of Michael and Irene's four children to work at the family dairy. She had a soft voice and whole-hearted smile, and was a mother to two children. Their names were tattooed onto one of her sinewy biceps, the other bicep reserved for a sprawling tree before a sizzling sunset in a turquoise sky. Annie's affection for animals set her apart from her parents and from most in her community.

Annie, for instance, named all the cows. "Every cow has her own personality," she told me, "but I'd say that most of them are gentle and friendly. They form friendships just like we do. Socializing is very important to them. I don't like the idea of identifying them just by their ear tag numbers, so when they're born, I write down names for all of them in my book, next to their numbers. No one knows their names except for me."

Miller milk plant workers Ken and Danielle, in contrast, viewed cows as just numbers. Both came from beef-cattle farming families, and neither had ever known anyone except Annie to name their cows. "Unless maybe," mused Danielle jokingly, "farmers call their cattle beasts Buttercup-one, Buttercup-two, Buttercup-three, until a hundred, just to say they name them!"

Danielle and Ken both believed that their "cattle beasts" did not recognize them. "If they're too stupid to recognize each other," Danielle reasoned, "how can they recognize people?"

Annie, in contrast, was confident that cows recognize one another, and also people. "I've noticed that when cows line up in stalls, they prefer lining up next to certain other cows. That shows they recognize each other and make friends—they like being next to their friends. Also, I'm not usually the one to milk them, but when I do milk them, I can tell they're surprised. That shows they recognize people."

I agreed with Annie. With every afternoon that I visited the cowshed, fewer cows stared at me, more of them started to ignore me. The rapid alteration in their reaction indicated that they recognized me as the same dawdling individual strolling among them every day.

To my delight, calves recognized me, too.

The seven youngest calves at the Miller farm had pink noses, inky eyes, and dappled designs of night skies suffused with white clouds.

Each calf was restricted to an individual five-by-six-foot enclosure called a hutch. The walls of the hutches were marked with long, dark, lightning-like streaks of excrement. The straw on their floors was stained by manure from the yellow of a banana to the brown of a potato peel. The knees and hooves of calves—and, in a few cases, even their bellies and necks and faces—were smudged and spotted with sticky dabs of feces. The calves looked dirty and unhappy.

The hutches stood in a row outside the cowshed. The first time I stopped by, all seven calves continued to lie idly in their lairs. As I walked to and fro in front of them, they stared at me warily, like I was trespassing on their terrain. Eventually, 307, an ebony-faced calf with a white triangle blazoned across her forehead, rose and stepped out into the small outdoor area at the front of her hutch. Her knees knobby, her eyes steady, she inched toward me tremulously. She sniffed my hand—then licked it. Surprised, I returned my hand to my pocket. Undeterred, she licked my pocket.

The other calves remained timid and placid that first day, but they became bold and brazen over the next days. They started calling out to me like flea market vendors, sauntering and sashaying to persuade their one visitor to arrive, or to return, or to stay. My favorite calf was 310 because she was the youngest and prettiest, her hair pattern that of a snow field sprinkled with pebbles, her eyelashes white around the left eye and black around the right. Even timid little 310 would step out of her hutch eagerly for a pat.

Calves were "more friendly and playful than puppies," as Michael himself said, and they craved companionship like sunflowers crave sunshine, but in their hutches they were permitted only flies for fellowship.

Hutches are the standard form of calf housing in the United States and Canada, but are less so in Europe. In 1998, the European Union banned the individual confinement of calves over eight weeks of age. The European directive "laying down minimum standards for the protection of calves" offered as explanation: "[I]t is recognized scientifically that calves should benefit from an environment corresponding to their needs as a herd-living species; whereas, for that reason, they should be reared in groups."

Organic dairy farms in Europe do not wait until calves are eight weeks old to house them in groups. They require that calves be housed in groups by no later than *one* week of age. In stark contrast, organic dairy farms in Canada (including the Millers') tend to house calves alone for a full three months. In the United States, it's worse: organic dairy farms are legally permitted to house calves alone until they're six months old—even though, at that age, a calf would fit into a hutch little better than a six-year-old child into a toddler's crib.

At *non*-organic dairy farms in the United States and Canada—often called "conventional" farms, because they're the norm, constituting all but one or two percent of agriculture—calf housing is entirely unregulated. Conventional farms can confine calves as long as they wish, in any manner they wish. When it comes to calves, organic farms in North America are a step up from conventional farms, but they're still a step down from conventional farms in Europe, and they're several steps down from organic farms in Europe.

"Damn you!" Irene cursed Michael one afternoon over lunch. "You promised me you would sell the farm, but you *still* haven't sold it! You're a *LIAR!*"

She stomped upstairs and slammed her bedroom door. Michael, imitating her demeanor if not her diction, strode out of the house and slammed the front door.

"The local paper we get here is pretty bad," Annie told me, looking up from her newspaper, "but I read it when I eat with my parents. You should, too. It's a life-saver."

Irene believed freedom from the farm could be won by eating Michael's soul. Michael believed freedom could be won by eschewing his house and spouse. Far from being the rural vacation I'd sought, the Miller farm was a land of altercation, pulsing and palpitating with tension. I regretted having committed to a two-week duration.

Like Michael, I whiled my hours away anywhere outside the house. It was during one of my whiled-away hours that I met Arthur the Artificial Inseminator.

Friendly, stocky, about forty, Arthur appeared dressed for a "Harry Potter Meets Agriculture" costume party. He wore gray coveralls and his left hand was enveloped in a bubble-gum pink glove that stretched all the way to his shoulder, his fingers clutching what looked to be a wand.

Arthur unplugged the wire electrifying the shit trainers. He then approached a cow Michael had marked for his attention with a blue string. He swept her tail aside. She jerked her head back, the movement forming the extent of her shackled resistance. Arthur thrust his arm into her rear. She blared and wrestled against her chain, striking the jagged-edged trainer above her repeatedly until it swung from side to side like a manic pendulum. Arthur remained unfazed—he'd unplugged the trainer wire earlier precisely because he'd anticipated the struggle. His arm surged farther, past his elbow, eventually entering "all the way into the rectum to straighten the cervix."

Arthur was too absorbed in the cow's rectum to notice that I was squirming. With his glove and wand, his manner alternated between that of a magician in the midst of a trick and a specialized technician giving a scientific demonstration. He seemed flattered to have an audience in me, and he described his rear-end activity with a cheerful, skipping-along joviality—even when the cow defecated.

Soupy brown torrents gushed out a gasp away from Arthur's face. I leapt back, but Arthur remained impressively unflustered, his hand still embroiled in the cow's defecating backside. He'd had more encounters with cow excrement than he could count, and they no longer held any power to repulse him.

Arthur waited for the deluge of dung to dwindle to a drip, then he penetrated the cow with his wand, which was embedded with a four-inch straw of frozen bull semen. When he extricated his arm, his glove was no longer pink but black. He flung the glove into the manure gutter, then plugged in the shit trainer wire he'd disconnected earlier. I told him I was both curious and scared to touch a shit trainer. "Should I touch one?" Arthur shook his head wildly, warning, "The power's *strong*!"

"This time, I'm doing just this one cow," he continued, "but I usually do more. If this cow doesn't get pregnant, I'll come back with more semen for her. I come here once or twice a week. I stay *very* busy. I have ten more dairy farms to service just this afternoon!"

Arthur showed me his agenda for the day on his phone. It was packed from morning to evening. But Arthur was in no hurry to get to his remaining farms; it was rare for him to have human company during his insemination endeavors, and he was loath to relinquish it. Bull semen is an enormous worldwide industry, he told me proudly, with its own corporations and corporate alliances, and acronyms of corporations and corporate alliances. He worked for Gencor (Genetics Corporation), which had just merged with EBI (Eastern Breeders Inc.) to create EastGen, which had then partnered with WestGen of British Columbia and the insemination center of Quebec to form an immense insemination entity called Semex (Semen Exchange).

"Canadian bulls have the best semen in the world," Arthur claimed patriotically.

Semex offers cow insemination every day of the week, every week of the year, everywhere in Canada, and also everywhere else in the world. The company consists of 1,600 bulls, 1,800 people, and semen distributors in 80 countries, ranging from Sweden to Sudan, Sri Lanka to Slovenia, India to Indonesia, and Australia to Argentina. Regarding itself as "world renown [*sic*] for delivering high quality bovine genetics," Semex describes its bulls like software products. They are "new releases" and they form parts of "series." They are alphabetized and codified in spreadsheet-like catalogues, each bull listed with a row of twenty numbers.

Artificial insemination has had a greater impact on the dairy industry than on any other area of animal agriculture, being used today in more than nine tenths of American and Canadian dairy cows. The stringent genetic selection that goes into artificial insemination has led to an extreme aggrandizement of the dairy cow's udder, which is today fleshy, busty, barely fitting between the back legs, veins popping—like a balloon about to burst. The average Holstein dairy cow produces about 20,000 pounds of milk per year.

Artificial insemination is not only unnatural, but harmful. Millions of dairy cows today all over the world are the descendants of only a few dozen bulls. This is equivalent to passing a complex, swirling pot of genes through a needle pinhole. Such narrow genetic limitation never occurs in nature, because the long-term survival of any species depends on genetic diversity. Also, the traits selected by insemination corporations are severely short-term, focusing on milk production at the expense of leg strength and structural balance. The dairy industry today deems milk to be more important than the cow, forgetting that there can be no milk without the cow.

"All children were to be begotten by artificial insemination . . . and brought up in public institutions," wrote George Orwell in *1984*. "One egg, one embryo, one adult—normality," wrote

Aldous Huxley in *Brave New World*. ". . . Making ninety-six human beings grow where only one grew before. Progress. . . . The principle of mass production at last applied to biology."

Orwell's and Huxley's dystopian predictions never came true for humanity, but they did for animal agriculture. This is surprising, given the fact that the prospects of artificial insemination were once viewed to be so dim that in a 400-page animal breeding textbook published in 1943, the practice was mentioned only once—in a single, cursory paragraph close to the end. "Artificial insemination is sometimes useful in overcoming sterility but has yet found only limited use," wrote Jay Lush in *Animal Breeding Plans*.

Only a few years later, American scientists discovered how to extract bull semen and impregnate cows with it. The initial challenge—how to keep semen fresh long enough for transport and use—was overcome by adding chemicals and antibiotics to the semen, and by freezing it.

Artificial insemination is today employed for different reasons in different areas of animal agriculture. In turkeys, it is used because male turkeys are genetically bred to reach a weight of 30 to 40 pounds, double that of females, and they cannot mount females without harming them. In fish, it is used because some fish species, such as trout and salmon, lay eggs only in gravel, and land-based fish farms find it easier to inseminate fish than to haul in gravel.

The rate at which artificial insemination is replacing natural reproduction is startling. In the one decade from 1991 to 2000, the proportion of American sows bred by artificial insemination exploded from less than 8 percent to nearly 70 percent (and today to more than 90 percent).

Artificial inseminators like Arthur, employed by corporations like Semex, regularly roam the countryside with tanks of frozen semen.

As I was striding past the hutches one evening, I noticed that the calves, normally quiet, were howling, bellowing to the full extent permitted by their lungs, their chorus shattering the still night air.

I went to the Miller house and asked Irene why. "They're bawling because they're hungry," she said. "They haven't eaten since the morning. I was supposed to feed them two hours ago."

"So . . . can you feed them?"

She shrugged. She arranged pots in the kitchen. She swigged her daily bottle of wine. She opened her mail. She informed me, yet again, of her myriad ideas for escaping the farm—becoming a translator, carving animal bones into pendants, selling art in Europe. "I don't care what I do or where I go," she concluded glumly. "I just want to leave."

Only when Irene could think of nothing more to say or do did she relieve the hunger of her charges. I realized then that catering to the calves and cows, themselves hutched and tethered, was, to Irene's mind, synonymous with starving her spirit. They were tying her to her barren life and preventing her from roaming the faraway gardens she wished to roam.

But they, too, wished to roam. Enough to escape.

Ten cows tore across the snow, pulverizing it into powdery flakes and cotton balls that rose in clouds around their feet. The black-and-white stampede was a photographer's fantasy, except for Michael—red, angry, out of breath, chasing them. "Help me bring them back in!" he yelled at me. "Move and stand in front of them!"

I did not.

"*MOOOVVVE!!!* STAND IN FRONT OF *THEMMM!!!*"

I jogged toward the cows, but haltingly, terrified of becoming a casualty of their stampede. In the cowshed, shackled by chains, the cows were still and silent like trees, and I felt safe among them. I did not envision their running any more than I would have envisioned trees running. But here, on this wide, white expanse of snow, the cows were barreling toward me with the weight and velocity of cars, brimming with the pent-up energy of their inactivity. One hit of the head, one kick of the leg—from just one of them—and I would be done for, stamped into the snow for eternity.

When the cows saw me blocking their path with my 105-pound frame, they slowed. Then . . . stopped. They were waiting for me to get out of the way, I realized, just as cars wait for pedestrians to cross the street. This was what Michael had in mind; he was able to catch up with them.

The cows were not milking cows but heifers, who had not yet given birth to their first calf. Every morning, when the cow shed was cleaned, the milking cows were moved to the heifer barn in order to ensure that their legs and tails did not get trapped in the churning manure gutter. To make room for the milking cows in the heifer barn, the heifers were shifted to the fenced area outside their barn. All the shifting was straightforward, routine as the sunrise, and there were never any accidents—except that morning.

That morning, as Michael was stepping out of the milk plant, he was shocked to discover that almost half of his heifers—ten of twenty-one—had walked out through the fence gate, which was mysteriously wide open. It had either been pushed open by the wind, or, more likely, unlatched by a long, wily cow tongue. He started chasing the heifers, and they bolted through the trees. He was annoyed, but not surprised, for he knew that cows enjoyed being outdoors, to the extent of even dreaming about it.

"All mammals dream," he'd told me earlier. "Dreams are a way for the mind to process things. Sometimes, you'll see a cow suddenly get up in her stall, or you'll see her moving her legs while she's lying down. That means she's dreaming. I think she's dreaming about being outdoors."

Dreaming or not, cold is no impediment to a cow, I learned. "Cows don't really get cold because they have their thick skin and thick hair," Michael had continued. "It's the same way with other farm animals. They don't like rain—they get sick if they're wet all the time—but they don't mind cold too much. Sheep have their wool for warmth. Chickens and turkeys have their feathers. Farm animals are fine outdoors for at least three-quarters of the year, even in places like Canada. Many are fine outdoors all year. Our calves are in hutches outdoors all year."

He was right; the heifers didn't even seem to notice the winter chill. Looking at them, I realized for the first time that farm animals *can* be outdoors for most of the year. Judging from their excited state, the heifers seemed to prefer the outdoors to the indoors no matter the weather.

With my help and his daughter Annie's help, Michael was eventually able to herd the heifers back through the gate. He locked the gate behind them.

Michael, who was stoically opposed to surprises, received not one, but two surprises, on his fifty-seventh birthday.

The first surprise was a snowstorm, and it forced Michael to spend most of the day at home with Irene. A snowstorm would have been unpleasant on any day, but it was especially unpleasant on his fifty-seventh birthday because the day marked his fourth missed deadline to sell the farm, and Irene was alternating between sulking and scolding.

The second surprise began with an urgent phone call from Annie. "Can you fix a pipe in my kitchen right now?" she implored her father. He would normally have pounced on any reason to escape Irene's tirades, but on this night, the tempest raging outside the house was more terrifying than that whirling within. "I'll fix your kitchen pipe tomorrow," he promised.

Annie insisted. He succumbed. She then asked to speak with me. After our brief conversation, I informed him that I would accompany him to her house.

Michael's car was buried under a hilly heap of snow like the fluffy white of a bride's dress. He managed to clear its windshield but not its windows. The drive to Annie's house felt like an action movie, for the opaque windows meant that to watch for other vehicles, we had to open our car doors and peek out—*with* the car still in motion.

We reached Annie's house trembling but alive, and Michael then learned what I already knew: there was no broken kitchen pipe. The pipe was merely a pretext Annie had invented to get her father to her house for the surprise birthday party she'd organized for him. Annie had asked me to accompany him because Irene had declined the honor. I'd agreed because I sought a change from my lackluster evening routine of trying to blend into the furniture as Michael and Irene fought, or else counting cobwebs and staring at the walls of my room.

Annie's house was small and simple, with daffodil-yellow kitchen walls. But on this stormy evening, the cheery color of the walls could not lighten the leaden mood of the five individuals Annie had convened around her kitchen table to pay their respects to her father. They were: her elder sister, her brother-in-law, a gardener, a car saleswoman, and Brick Roberts.

Brick looked as farmerly as a man could. Standing at medium height, he had a flushed face, and his belly pouted like a pear even as his back tapered like a pepper. His incessant chatter, his scarlet sweater, and his bird's-nest beard made him reminiscent of Santa Claus. The likeness became complete when he handed

Michael two bottles of wine he'd tucked into long gray socks to imitate reindeer antlers.

Brick stood very close to me as we spoke, his billowing belly grazing mine, his aura of cigarettes accosting my senses. I could see the white hairs in his charcoal-gray moustache and beard, the bloodshot streaks in his dark blue eyes, the mottled veins in his cheeks.

"What do you do?" I asked him.

"I have a few hens for eggs," he replied.

"How many hens do you have?"

"A few."

"Like how many?"

"A few."

"I see," I said, though I didn't see at all. "Can I visit your hen farm?"

"Hmm. . . ." Brick frowned at length. He eventually reached into his pocket and handed me his business card.

"I'm *very* surprised Brick gave you his card," Michael told me when we returned to his house. "Farms like his never let people in."

"Why?"

"Why! Why do you *think*? Because the egg-laying hens are in cages. Cages are extremely inhumane. You can call the number on Brick's card if you want, but you'll be wasting your time. You'll get a lot of bullshit, and Brick won't let you see his hens. I've known Brick for twenty-five years—the length of your whole life—and *I've* never seen his hens."

I would call Brick soon, I decided.

"Your calves are cute," I complimented Michael and Irene over dinner the next evening, in an attempt to make conversation.

Michael set down his fork and knife, his face reddening. "Calves are not '*cute*,'" he said with a glower. "Animals convert what we feed them into products we can use. They digest food we can't to make something for us. As long as we give them food and water, we're doing everything we need to do. Animals are never '*cute*.' We can't think of them like that. If you don't understand something so basic, you shouldn't have come here."

Whether or not I should have come, it was increasingly evident that I should not remain. I packed up and departed the next morning, shearing my residence with the Millers from the two weeks originally planned to nine days. Every day with the Millers had dispelled my sing-song notion that life on an organic farm is like *Little House on the Prairie*. Michael and Irene were not married happily. They did not dine organically. They did not view their animals favorably. They did not tend to them compassionately. They did not even feed their calves regularly.

The Miller mindset that we owe animals no more than food, water, and shelter is flawed. After all, if we give our cats and dogs food, water, and shelter, but keep them chained by the neck under electric trainers, that's considered inhumane and unacceptable. The physical needs of food, water, and shelter are only basic, beginning needs. Another important physical need shared by all animals is movement. The word "animal" shares a root with "*anima*tion" because it is animation—the affinity and proclivity for movement—that differentiates an animal from a plant, the latter of which is *plant*ed in one place. For social animals, there is also a need to interact with others, and for many animals, including cows, who have evolved as grass-eaters, there is a need to graze outdoors.

The Miller decision to go organic had had more to do with profit incentives than philosophy. Even had it not, however, Canadian organic standards are flimsy. Canada's guiding organic document states, "Use natural methods of reproduction; however, artificial insemination is permitted," and "Except for lactating [dairy] cows, the continuous tethering of livestock is not permitted."

Consequently, both artificial insemination and the "continuous tethering" of dairy cows were the mode of practice at the Miller dairy farm. A history of organic aids understanding.

The word "organic" started being applied to food in the 1940s, coined in reaction to the creation of artificial fertilizers and pesticides. The concept of organic didn't catch on until the 1960s, however, and it wasn't until the 1970s that organic started being certified by independent certifiers. The United States Department of Agriculture (USDA) had little to do with organic at that stage, getting involved only long after that.

The USDA initiated the National Organic Program in 2000 by publishing a proposed rule about it in the Federal Register. After a review process—the proposed rule received 40,000 comments from the public—the USDA published a final rule on the National Organic Program, which went into implementation in 2002. The comments that the USDA received on its proposed rule highlight consumer concerns about organic animal farms.

The USDA's proposed rule mentioned "access to shade, shelter, exercise areas, fresh air, and direct sunlight," but it did not directly mention "access to the outdoors." Commenters complained that USDA's language suffered from a lack of clarity. One commenter expounded astutely that "the requirement that animals receive direct sunlight could be interpreted to simply require windows in livestock confinement facilities." The USDA conceded by adding an "access to the outdoors" provision to its final rule.

The public noted another issue with the proposed rule: "access to pasture versus pasture-based." The proposed rule mentioned "access to pasture," but commenters stated that this term did not "sufficiently characterize the relationship that should exist between ruminants and the land they graze." The term "access to pasture" suggests being able to access pasture, whereas "pasture-based" connotes living on pasture, or at least grazing on it at length.

The National Organic Standards Board (NOSB)—a federal advisory committee of fifteen individuals appointed by the Secretary of Agriculture to advise on the National Organic Program—agreed with commenters. The USDA still declined to alter the language in the final rule to "pasture-based."

The proposed rule also contained a provision for "stage of production." Commenters worried that dairy producers could argue that lactation is a "stage of production," and wrongly use that provision to deny dairy cows access to pasture. The NOSB again agreed with commenters, but the USDA again declined to make the change. NOSB and commenter fears were precisely realized when organic dairy producers denied cows access to pasture.

The NOSB, however, persisted in both the "access to pasture" and "stage of production" provisions, voicing their recommendations again in 2005. "Lactation of dairy animals is not a stage of production that justifies confinement and keeping animals off pasture," they argued. This time, the USDA heeded their counsel and defined "stage of production" to exclude lactating dairy cows. This new provision became part of USDA's final rule on Organic Access to Pasture, which went into implementation five years later, in 2010. The practice of continuously tethering cows by neck chains, as at the Miller dairy farm in Canada, effectively became prohibited on American organic dairy farms in 2010.

This updated "stage of production" provision regarding dairy cows created one of the few differences in organic regulations between the United States and Canada. Organic standards in the two countries have almost everything else in common, including the fact that they mandate access to pasture for a minimum of 120 days of the year. This translates to one out of every three days. At many organic farms, including the Miller farm, the minimum is also treated as the maximum. Miller dairy cows stepped outside on exactly 120 days of the year, not one more or one less.

I returned to the Miller farm in the fall to witness the cows' outdoor days.

I was delighted to see that the farm landscape was a meadow—melted, mellowed, unrecognizable from the white desert it was in winter. The cows' lives were also unrecognizable. They strolled on fields of dewy green grass lined with orange pumpkins and scarlet carpets of leaves. The panoramic sight before me meant that I was wrong before my first visit, and I was wrong also after it.

I was wrong before my first visit to assume organic meant excellent, and I was wrong after it to infer, as Miller milk plant workers Ken and Danielle did, that "There's no difference between organic and non-organic farms in terms of how you treat animals."

Had the Miller farm *not* been organic, the dairy cows would have spent not two-thirds of the year, but *all* of the year in stalls and chains. As such, the 120 days outdoor requirement did form a crucial difference between organic and conventional farms.

To be more meaningful, the organic outdoor requirement for animals should be stronger, as it is in Europe. Austrian organic standards require animals to be outdoors for at least 180 days of the year: every other day. Swiss organic standards require animals to have access to pasture on at least twenty-six days per month in the summer, and thirteen days per month in the winter.

If the Miller farm was organic, what was a conventional farm, I wondered.

Brick Roberts would know.

CHAPTER TWO

EGG AGONY

CANNIBALISM AMONG CRIMSON COMBS

The aluminum walls of Brick Roberts's office were plastered with papers, all of them crinkled and yellowed except one: an outdated calendar of a topless woman cavorting on a beach. His feet up on his messy desk, the soles of his big, brown boots facing me, a steaming mug of black coffee in his left hand and a dwindling cigarette in his right hand, Brick opined on everything. And more.

"Laws against drinking and driving ruin lives out here in the country. 'Cause people can't drink outside and drive home, they drink at home. Then husbands beat their wives. Wives beat their husbands. People commit suicide! We live in a Communist country. It's not right I can't smoke in restaurants—it's Communist. People have lost all common sense. . . .

"I always carry a lighter. It's not just for cigarettes; I burn my trash with it, too, 'cause I don't want it ending up in a landfill. There's too much stuff in the world to fit into landfills. I can let ya burn some of my trash, too, if yer interested. I'm an environmentalist. And I'm a liberal. I don't like war. . . .

"I'm scared of the dentist, but ya know what's even worse than getting a root canal done? Girl stuff. Like shopping. I buy twenty-four pairs of socks when I go shopping, so I never have to buy socks again. I hate shopping. But I like girls. My favorite city in the world's Vegas. I've been to Vegas thirty-five or maybe forty times. I got a bitch of a hangover every time I went there. I chased wild women and gambled and did all that good stuff when I was younger. I don't see anything wrong with liking young women. . . ."

Perhaps it was because I was a young woman that Brick had agreed to my visit. He'd taken me for younger than I was, though, his jaw dropping to learn that I was not eighteen, as he'd assumed at Michael Miller's birthday party. It had taken me four phone calls to arrange my visit with him.

"I'm real busy today," Brick had sighed over the phone two days earlier. "Gimme a ding tomorrow."

"Bad day today," he'd grumbled the day before. "Gimme a ding tomorrow."

"Good day today," he'd chirped earlier that morning. "Gimme a ding later to see what I'm doing."

"I might have a little time for ya," he'd ventured during my last ding.

I'd hurried to Brick's office before he could change his mind, and now, after listening to his miscellaneous opinions for more than an hour, after even being apprised of the intricacies of his socks-purchasing preferences, I asked him for a tour of his egg farm.

"Sure thing," Brick said.

We climbed into his dusty jeep and drove past long rows of spruce and cedar trees, and snow-cloaked fields that in the summer grew corn for Brick's birds and soybeans and wheat for the market as cash crops. After a slow, half-hour meander around the property, Brick drove us back to his office. I felt a surge of disappointment that he hadn't stopped at his egg farm.

I didn't know it then, but fourth-generation farmer Brick *had* stopped at his egg farm. He'd also started there. Brick's egg-laying hens were housed directly behind the papered-up aluminum wall behind his desk, in the endless, windowless section of the rust-brown building we were in. We'd been at the egg farm the whole time.

"You can call the number on Brick's card if you want," I remembered Michael Miller telling me, "but you'll be wasting your time. You'll get a lot of bullshit, and Brick won't let you see his hens." Michael had been friends with Brick for a quarter-century, and *he'd* never seen Brick's hens. I'd known Brick for only a few hours, in contrast; what were the chances I'd succeed where Michael had failed?

Time flew by as Brick and I continued to talk. We were in most manners opposites—I was quiet and reserved; he was warm and funny, bellicose and opinionated—but conversation between us flowed like water. Before I knew it, it was five o'clock, and Brick was hauling his feet off his desk to return home. His secrecy about his egg farm had only enhanced my curiosity about it, however, and I made a last-ditch effort to see it. "Do you like animals?" I asked him.

"I sure do," Brick answered, his eyebrows ascending in surprise, for the question was unrelated to the story he'd just recounted, of his chagrin at having failed the sixth grade. "I make a living off them. . . . Ya do what ya have to do to make a living."

"Can I see your hens?"

Brick's expression became illegible, the creases of his countenance ordering themselves into a closed configuration underscored by the straight streak of lips finally united. He sat perfectly still for a moment, then he pushed back his leather chair, jumped to his feet, bounded out of his office, and pushed open the adjacent door.

The *reek*.

It was rancid and repulsive, spitting us out, but also swallowing us into its monstrous metal machine belly, whose bowels were thousands of cages arranged in three endless columns. Like the shelves of a storeroom, each column was horizontally divided into four rows of cages. Each row consisted of two sets of adjoining cages, just as grocery store shelves are stocked from both sides of the aisle. Each cage was the size of a microwave but confined four or five hens. There were 13,000 hens in the factory.

The hens' crimson combs, august like crowns, contrasted against the encompassing slabs of steel like painted red lips against a desiccated face. All the hens were crowned with combs, but not all wore the auburn feathers of their Rhode Island Red breed. Many wore only a meager under-layer of down feathers, which looked like a coat filling without a coat—impoverished, unkempt. Other hens were entirely pink and bald across the breast and back, and especially across the neck. When destitute of all but a few spike-like plumes, the neck was frighteningly thin—the diameter of a quarter—and resembled a leafless seedling languishing in a pot its owner had forgotten to pick up and place in the sun.

And there was no sun. There were no windows, and sunlight could not sidle in even through exhaust fans, for their blades whirled under a carpet-thick moss of dust, dirt, and feathers. Yellow bulbs dangled on long wires from the ceiling, but they were dim and dwarfish and deficient, their jaundiced glow an insufficient substitute for sunlight.

Cages were the epitome of cold, steely efficiency. Hens drank water from a dripper at the back of their cage, and they ate yellow-brown pellets off a feed belt at the outside front of their cage. They lived underneath a manure belt that sagged under the waste of the hens above, and they stood on a wire floor that was slanted so their eggs would roll automatically onto the egg belt running parallel to the feed belt outside their cage. Each cage was a small, self-sufficient, automated black box.

Such cages are called battery cages because of their vast numbers, as in a "battery" of tests. The term "battery" also has military implications—an "artillery battery" refers to a row of guns or missiles. The term's association with war is appropriate because, first, battery cages came into commercial use in North America after World War II, and, second, the very concept behind battery cages is control and incarceration.

Before the invention of battery cages, hens lived in small backyard flocks. Cages were developed to make it easier for egg producers to identify and eliminate "passenger" hens—those hens whose egg output did not financially justify their continued existence. The materials required to construct cages were expensive, however, and the egg industry soon realized that it could save money by confining more than one hen to a cage, even if this meant being unable to get rid of passenger hens. The number of hens per cage consequently increased to two, and then three, and, today, four to ten. As for cages, they are today stacked anywhere from four to eight levels.

Seeing Brick and me striding down an aisle, the hens squawked wildly, their sharp, shrill shrieks amplifying as they reverberated off the metal. They stampeded all of one step to the dark rears of their cages, where they trampled one another in an effort to conceal themselves.

I'd judged Michael Miller's dairy farm severely—with its neck tethers, cow trainers, and manure gutters—but it seemed like a slice of heaven in comparison to Brick Roberts's stacked, stinking warehouse. Michael's cows at least stepped out on pasture one out of three days of the year; Brick's hens, in contrast, would live in cramped cages until the day they were slaughtered. Michael was right to describe Brick's egg operation as "extremely inhumane."

Brick wanted us to leave as soon as we entered. I wanted to return as soon as I left. I wanted—*needed*—to understand what I'd just seen.

I lived close to three hours away from Brick, and so repeated day trips were not a viable option. As I saw it, the only way for me to learn more about Brick's profession would be to live with him in his house. The prospect made me nervous, though, for I hardly knew him—except that he liked young women—and my stay with the Millers had been disastrous.

I mustered up the courage to call Brick two weeks later.

He no longer remembered me, speaking to me as skeptically as if I were a telemarketer.

I reminded him of our hours of conversation in his office. "Do you remember?!"

"Hmm, kinda. . . ." he replied.

Given Brick's hazy memory, I knew I was being pushy and presumptuous, but I still asked, my voice a whisper, "Can I come stay with you?"

I thought Brick would ask me why; anyone else would. He didn't. It was almost as if he'd expected the request. "I'm real busy this week," he sighed. "Gimme a ding next week."

Next week: "My wife's at her uncle's funeral. Gimme a ding next week."

Next week: "I'm having trouble with the chimney. But, sure thing, ya can come on over in a week."

In a week, when I arrived at the bus stop in his village, I was afraid Brick would fail to recognize me. Instead, he enveloped me in the sort of squeeze-embrace reserved for family members. Lifting my suitcase as effortlessly as if it were a pillow, he invited me to dine with him at the only restaurant within three miles of his house: The Truck Stop.

At the appropriately named diner, among heavy customers and hairy owners, in a booth with cracked seats, Brick ate almost as much as Michael Miller's dairy cows. He devoured two mugs of hot chocolate, a bowl of ham and vegetable soup, and a staggering platter of ribs, wings, and fries, over which he first dragged the salt-shaker methodically, meticulously, like a satellite.

I'd never seen anyone eat as much food or as much salt in one sitting. "I know salt's not good for ya," Brick said with a shrug when he noticed my open-mouthed stare. After he'd eaten every morsel on his plate, Brick also dispatched the two lollipops that appeared with the check. (He had his and eyed mine; I offered it to him, and he accepted it gladly.) He then drove us home in his jeep. His house was impressive, to my surprise—I'd been nervously prepared to be slumming it, given the derelict state of his office. Brick lived in a two-story, red-brick residence with a three-truck garage. His wife greeted us at the door.

Jane Roberts was blond, round, and, as Michael Miller once described her, a "nice lady," with a motherly manner and tranquil tone. Whereas Brick's face was leathered and furrowed like a canyon—its ridges carved by caffeine, its trenches cut by nicotine—Jane's was traced with a faint patchwork of cobweb-thin threads. Brick and Jane, by their appearance and personalities, were an unlikely couple, and their first meeting had also been unlikely. They'd "clicked" when Brick had hired Jane and her friends to load his egg-laying hens onto trucks for slaughter. A few years later, after Brick had pared down his Vegas "wild women" chases, he and Jane, both thirty, had read their vows before four guests.

Jane was a sprinkle of sugar on Brick's salt, though she ate little sugar, and also little salt. Even when she dined at The Truck Stop, she ordered a salad, usually a Caesar salad. In recent months, however, she was frequenting The Truck Stop less and less because one of her evenings every week was spent at a Weight Watchers meeting. She'd shed thirty pounds since joining the program. Brick knew Weight Watchers only as "fat camp," though he took care to warn, "Don't go telling Jane I call it fat camp!"

Jane hardly ever came across a vegetable she didn't like; Brick hardly ever came across one he did like, finding that he could truly tolerate only two—celery, occasionally, and potatoes, frequently, preferably in the form of fries. Brick and Jane's divergence on diet was thus considerable, but it was still trivial when compared to their divergence on cigarettes, coffee, and alcohol.

Jane did not smoke, stating "People around here smoke *way* too much"; Brick held himself to a daily quota of fifty cigarettes. Jane occasionally sipped a cup of coffee softened with milk; Brick swilled a dozen mugs of black coffee every day, even though the caffeine gave him sleepless nights and insomniac eyes, veined with red streaks crisscrossing like tree branches. Jane drank beer and wine socially; Brick enjoyed alcohol immensely, storing hundreds of bottles of red and white wine in the wine cellar in his basement.

Brick and Jane shared their home with Dexter, an obese, drowsy cat, and Lizzie, Brick's black, barking darling. "The most important job around here is loving Lizzie," Brick told me, patting her head. "Lizzie's always gotta be happy." It was strange and ironic to see Brick, who caged 13,000 hens, dote over his dog more than his sons, Nick and Will.

Nick was a twenty-three-year-old with a scruffy beard and sturdy build—a young replica of Brick. Will was a tall, sleepy twenty-year-old who wore, even at home, an orange baseball cap patterned with silhouettes of curvaceous women.

"Nick and Will will be yer country tour guides," Brick offered me, without checking with them first. "They'll do anything ya want. They'll take ya anywhere ya want."

Will wanted to take me to a friend's hot tub later that night. "Do you have a boyfriend?" he asked.

"I don't," I replied.

"I'm single," he shared. "Sorta, kinda."

Nick asked me the same question separately, also sharing his single status.

Brick saved his favorite possession for the end of the house tour. Standing at the foot of the stairs in the living room, he reached inside an ornamental pot and pulled out the family heirloom: a rifle. Jane's mother had, over the span of her life, deployed it to kill squirrels in the attic, but its present purpose was different, darker.

"I don't believe in locking the front door," Brick said. "If someone comes in who's not supposed to be here, that's what the gun's for. Use it if ya ever see a stranger in the house." Brick twirled the long, loaded, rusted weapon around his finger like a cheerleading baton; I stepped aside discreetly.

His gaze grew glassy and his voice became marshmallow-soft as his thoughts turned from the prospect of shooting strangers to the pleasures of killing animals. "I have a deer in the freezer," he said. "Ya can have some of it and take some back with ya for yer family."

I demurred.

Brick continued that he sometimes hunted alone and at other times he hunted with Will, who'd slayed his first stag on a blustery winter day six years ago, the event commemorated with a framed photo in the kitchen. His father's arm around his shoulders, Will, then fourteen, wearing a shy, self-conscious smile, propped the stag's head up for the picture by gripping his antlers. To the onlooker, however, holding the stag's head up seemed hardly necessary, for, if not for some snow stained scarlet underneath him, the young stag looked vibrantly alive, his dark eyes staring straight into the camera.

Since that first kill, hunting had transformed for Will from a pastime to a passion, and even almost a profession. Now, however, it was coyotes, not deer, he pursued, because "If you cut off their ears and take them in to the township to prove you hunted one, they'll give you a hundred dollars per coyote."

Ear-cutting was the primary exertion required in hunting, at least on Will's part. His pack of six hunting dogs—purchased for a total of a thousand dollars, some from the United States—were outfitted with GPS satellite collars. The collars appeared on the screen in Will's truck as dots. Seated comfortably behind the wheel, a bottle of beer in one hand and a cigarette in the other, his rifle resting on the passenger seat, Will merely followed the dots like he would any other GPS directions.

Replacing the gun in the floral pot, Brick warned me, "Don't tell anyone we have a gun. We're not allowed to have one. But if ya want to come hunting with me sometime, I'll let ya shoot, too."

I felt like I'd fallen into a foreign country. Farm country *was* foreign country, with its own customs and rituals, vocations and avocations.

I awoke the next morning with a sore back. My bed was lumpy and my pillows had a stuffing as splintered as breadcrumbs.

It was among pickup trucks, a jeep, a snowmobile, kayaks, chairs, hockey sticks, a beer keg, and bottles and cans of beer and soda that Brick breakfasted every morning, seated at a picnic bench somehow also wedged into the garage. "I can't start my day without a pack, and Jane doesn't allow smoking in the house," he explained to me. "What do ya want for breakfast? Beer? Or coffee?"

I declined both, opting instead for a banana and orange, to Brick's befuddlement.

Brick, who clutched in one hand a cigarette and in the other hand a beer mug brimming with coffee, was wearing the same denim shirt he'd worn the previous night and also during the afternoon I'd spent at his office. The shirt, a stiff medium-blue many years ago, was now pale, threadbare, and frayed, smudged with yellow sweat stains around the armpits and assorted splotches elsewhere. It was still in good shape, however, when compared with Brick's navy-blue pants: torn in the center of the seat, they offered an eyeful of polka-dot boxers. I never knew where to look when talking to Brick.

"Let's go to the office and see if Paul has something for ya to do," Brick said after he'd chugged a couple mugs of coffee.

Paul was a stout, square-jawed man who'd worked for the Robertses since he was fourteen, when Brick's father had hired him to work after school and in the summers. Brick, then eighteen, soon lost his father to a fatal accident, but he gained Paul, who soon became his "best friend" and the "closest thing to a brother" that he, an only child, had ever had. Today, at fifty, Paul vacationed with the Robertses annually (usually in the Caribbean), had lunch prepared by Jane daily, and was "Uncle Paul" to Brick and Jane's children permanently. Paul's responsibility was the egg-laying hen factory.

He operated it from an expansive switchboard. The switchboard's knobs and buttons, when pushed or pulled, fed hens, crated eggs, and dumped droppings. The switchboard did everything, in fact, except extract dead hens from cages, which was consequently Paul's first chore every morning. I helped him with it that morning, even though it was easier that morning than almost any other of the year. The hens were a fresh flock, new stock, impounded into the equipment only a week earlier, clothed still in a complete coat of quills. They had not yet pecked one another to the point of naked pink patches, unlike the batch of hens I'd seen the previous time.

Those hens were slaughtered two weeks prior to my arrival, at eighteen months of age, as is the norm in the egg industry. Most of their body parts—their back, neck, and legs—were shipped to China, forming part of an international, complex, uncontrollable food chain. The rest of their bodies were sold to McDonald's to make nuggets. (Dairy farmer Michael Miller's cows, too, after being sold at a livestock auction, were bought primarily by McDonald's to form hamburgers, Michael believed. Organic milk is sold at a premium, but, ironically, upon slaughter, organically raised dairy cows become part of the cheapest food in the country.)

After manipulating some knobs and buttons at the switchboard, Paul strode down the first aisle of cages. In his haste, he neglected to notice a glaringly visible dead hen. I indicated the bloodied, muddied carcass, covering half a cage floor like a carpet. Paul turned back and extricated it from the wiring. As he did, the living hens in the cage turned into a quartet, shrieking an off-key, melancholy melody.

Shuddering at the song, I lurched onward ahead of Paul. Only a few cages down, I spotted a second perished hen. Her head, which was a bloodless gray color, dangled through the tiny hinge gap between the door and the wall of her cage.

"It hanged itself," Paul said, disentangling the head from the door. "It probably stuck its head out 'cause it was bored, and it got stuck by accident. They always die when they get stuck. If it's a really small space, they choke, like this one did. If it's a bigger space, they don't choke, but they starve, 'cause they can't get to the food and water anymore. Hanging is one of our three big causes of death."

Hanging? I'd never heard of animal hangings before, and I couldn't believe such a form of animal death existed. But I had to believe it; just a few cages down, another hen had also "hanged" herself. The deaths of hanged hens could have been prevented if Paul had bothered to notice that they were trapped in the wire. Instead, they suffered every minute of the week that it took them to slowly starve and dehydrate to death.

"Another cause of death is putting them in the cages," Paul continued. His tone was as light and breezy as if we were strolling through an apple orchard. "The crew puts them in real fast, and the hens sometimes break a leg or a wing. But they don't die from the injury. They die 'cause they can't get up to eat or drink when they're injured."

Paul added a third carcass to the two in his hands. "But our number one cause of death is blowout," he said. "Blowout means the insides get pushed out of the body. It happens when the hens get older, when they've laid a lot of eggs."

Blowout is caused by hens laying eggs that are too large, and by starting to lay eggs at too early an age—two genetic traits that the egg industry selects for obsessively, because consumers prefer large eggs and because early-laying birds are more cost-efficient. Although blowout is primarily genetic, the death from it is environmental. Other hens in the cage peck the blowout hen's bright red and pink insides, leading to her eventual, tortured death.

Paul and I had walked only a little farther when Brick appeared suddenly behind us. Brick requested Paul's assistance with something in his office. Paul agreed and began trailing Brick to the door, concluding the dead-hen hunt before he and I had finished with even a tenth of the cages. In just those first cages, however—in mere minutes of searching, even though the hens were very young—we'd found three dead hens.

"I didn't notice any of these hens you saw," Paul told me. "You're real good at spotting them!"

I tried to smile at the compliment.

Later in the afternoon, Paul would retrieve three more dead hens, but this achievement would be a testament more to hen mortality than his perspicacity, for Paul was not good at spotting dead hens. In part, this was because he hurtled past the cages indifferently, and in part, it was because he glanced at only the second and third levels of the four floors of cages.

Scanning ground-floor cages would have required stopping, stooping, and squinting, since the cages were dark and shadowed like underground caves due to their distance from the ceiling bulbs. The highest floor of cages suffered from the opposite problem. The cages were illuminated but too elevated, well above eye level, their roof close to the ten-foot-high ceiling. This meant that Paul only ever peeked into them when he sniffed "something rotting." This in turn meant that—because of the Olympic-level competition for odor—dead hens on the fourth floor of cages putrefied for weeks.

The task of extracting fourth-floor hens required a form of rock-climbing, and Paul showed me how he did it. He clutched the third-floor feed belt and lifted a boot onto the second-floor feed belt, then he lugged himself up to peer into the fourth-floor cages. But he had no more than a few seconds to scan the cages for decomposing carcasses, for the hens panicked at the sight of his wide, light-blocking body. They flapped their wings frantically, hysterically, setting off a blinding, tear-inducing storm of dust and feathers.

"Now *you* do it," Paul coughed at me as he descended, heaving at the exertion.

Shoot. I shifted several steps away, in order to not terrify the same hens Paul had just terrified. Then, gripping the third-floor feed belt, I stepped onto the second-floor feed belt, and I yanked myself up as Paul had, until my eyes looked into a fourth-floor cage. The hens before me started trampolining on each other's backs like jack-in-the-boxes, in a frenzied attempt to escape.

But there was nowhere for them to go. Their entire life consisted of a rusty metal box of bars, walls, and wires. For the first time, I felt like a monster for being there.

I liked Paul almost as much as I liked Brick. He was polite, patient, talkative, and trusting, to a naïve degree even. Brick's wife, Jane, was right to describe him as "a big kid." I'd taken to following Paul around the egg factory, and was pleased to see that he seemed to enjoy my company as much as I did his. Strangely, our friendship was cemented by our earlier scavenger hunt for dead hens.

Paul turned three knobs at the switchboard (one knob for each column of cages), and the dozen manure belts (one belt for each row of cages) started pulsing and palpitating like airport baggage belts, their baggage a slimy pie of hen droppings. The belts conveyed the droppings to the back of the building, where steel blades scratched them off onto a central belt below. The central belt then rose steeply like a roller coaster, darted through a narrow tunnel high in the wall, and dumped its freight of feces into a deep outdoor pit with a storage capacity of 20,000 cubic feet of waste.

Paul's role in the ride was maintenance man. Perched on a three-step stool at the back of the factory, a small windshield wiper–like implement in his hand, he cleaned the steel blades periodically so that they could scrape more efficiently.

I'd noticed it from the first minute I'd first glanced at the hens, but, standing still, I couldn't stop staring at the hens' beaks. Their beaks were blunt stumps, the upper portion without a tip, the lower projecting like a pouting lower lip, giving the impression, odd but fitting, that the hens were sulking. Adding to their downcast appearance was a crusty black scab, found on about one beak in five. It was dried blood, spilled because Brick had—for no reason—had his hens de-beaked relatively late, at eight weeks of age, when their beaks were fully developed, fully sensitive.

De-beaking involves slicing off the beak with a hot blade in a machine called a de-beaker. The procedure is deemed to be so painful that it is banned in Switzerland, Sweden, Norway, Finland, and Denmark. The bans have been passed in deference to scientific evidence and in indifference to the egg industries of the United States and Canada, which consider de-beaking customary, almost mandatory. They view it as a kind of cancer cure, the cancer being hen cannibalism. Like other cancers, hen cannibalism is diagnosed in the United States and Canada as requiring a diversified approach in its prevention and treatment. De-beaking is merely the first step; next steps are cage-walling and low lighting.

"If hens in one cage start pecking each other," Brick told me, "hens in other cages start copying them, until they're *all* pecking each other. The buggers start killing each other, and sometimes they also eat each other. They become cannibals. Cannibalism's like a disease. It spreads real fast and can become a real big problem. We don't want that. That's why the sides of the cages are walls, not bars. That way, hens can't look into the cages on their right and left. They can't see what other hens are doing, and they can't copy them. They can look out only in front of 'em and behind 'em. . . . Lighting is real important, too. We keep the lights low so the hens are less active. If they got more light, they'd want to be more active, and they'd peck each other more. They're nasty little fuckers."

Cannibalism confounds the egg industry because it is influenced by both a hen's environment and genetics. Under natural living conditions, hens prefer to move away from sources of fear or frustration; in cages, they are unable to move away, and so they attack.

The genetic component to cannibalism means that it is not only caged hens but also cage-free hens who peck one another, sometimes to death. At the start of the twentieth century, hens, like other birds, used to "brood." They'd sit on their eggs in their nests and protect them from intruders, an avian maternal instinct. But the egg industry sought a higher egg output, and it began to genetically select for it by retaining hens with high egg output and eliminating those with low output. The single-minded efforts resulted in hens laying 150 eggs a year by the middle of the century and 300 eggs a year by the end of the century.

But egg output and brooding instinct, called "broodiness," are inversely related. Present-day hens lay more eggs than any hens who have ever lived, but they have minimal mothering capability. They're aggressive creatures. Hen genetics, when combined with cage confinement, leave Brick and the rest of the egg industry to resort to strategies such as de-beaking, light-dimming, and wall-erecting to combat the contagion of cannibalism.

De-beaking poses a catch-22. The procedure reduces pecking not merely by breaking the beak into a blunter instrument, but by turning it into a locus of pain. The hen becomes less inclined to use her beak for the same reason that a person is less inclined to punch with a wounded hand. But although de-beaking is painful, the consequence of leaving the beak alone can be cannibalism—more painful still.

One day of my stay, Jane Roberts took me to the flea market in her truck. We strolled among tables piled with tangerines and cauliflowers, jackets and clay bowls, tablecloths and storybooks. We encountered several people she knew and she introduced me to them as her "friend," to my flattered delight.

"I have a sister who's eighteen years older than me," Jane told me. "She married a butcher and learned the trade from him. When he died, she continued with it. I was twelve then, and I helped her, because she wanted me to. I didn't kill the cows, but I helped her cut them up. That's when I developed an aversion to beef, and to lamb. Later, I also developed an aversion to pork, when I worked at a pig farm, doing sow farrowing. But I didn't want my kids to have my aversions, so I cooked everything for them growing up. I still cook it for them; I just don't eat it myself. I can't imagine ever eating beef, pork, or lamb again. I eat vegetarian mostly. You may be interested to know, Michael and Irene Miller became vegetarian fifteen years ago. They're not vegetarian anymore, and I don't know how long they stayed vegetarian, but they were then."

Michael and Irene Miller—vegetarians? Jane, mostly vegetarian? Meat producers shunning meat? Who would have thought? Certainly not me—vegetarian myself.

When I was twenty, I came across undercover videos of factory farms online and decided to leave animals off my plate. The decision was significant because, until then, meat had formed my primary sustenance. I'd eaten animals every day since I was a child, without ever thinking of them as animals. The only stretch of time I'd been without meat was the five days of my ill-fated freshman-year hiking trip; I'd dealt with the absence of meat by stopping all food consumption period.

At twenty, when I anxiously decided to remove animals from my plate, the array of foods available to me did not decrease, but increase, to my surprise. I replaced the few kinds of animals I ate with dozens of kinds of plants. I loved the bright colors and healthful effects of my new plant-based meals.

When I'd watched undercover videos in university, I'd wondered whether they were exaggerating, emphasizing the exception rather than the rule. But they were not, I now saw—now that I was *in* the sorts of facilities depicted in the videos.

"When I visited a large egg layer operation, and saw old hens that had reached the end of their productive life, I was horrified," said Dr. Temple Grandin, a well-known animal expert and scientist, in a presentation to the National Institute of Animal Agriculture. "[Egg-laying hens] were nervous wrecks that had beaten off half their feathers by constant flapping against the cage. . . . The more I learned about the egg industry the more disgusted I got. Some of the practices that had become 'normal' for this industry were overt cruelty. Bad had become normal. Egg producers had become desensitized to suffering. There is a point where economics alone must not be the sole justification for an animal production practice. When the egg producers asked me if I wanted cheap eggs, I replied, 'Would you want to buy a shirt if it was five dollars cheaper and made by child slaves?' Hens are not human but research clearly shows that they feel pain and can suffer."

Europe agrees with Temple Grandin. Battery cages became illegal throughout the European Union on the first day of 2012. Across the Atlantic Ocean in North America, in contrast, more than 300 million egg-laying hens—more than nineteen of every twenty of them—continue to dwell in battery cages.

Jane Roberts feared that this circumstance would not last long, that the United States and Canada would soon follow in Europe's footsteps. "It hasn't happened yet in Canada," she said solemnly, "but it's already started happening in some American states. They don't allow cages anymore."

California, Michigan, and Ohio have passed legislation to phase out battery cages.

Eggs are flooded with as many misconceptions as labels. First among them is egg color: people view brown eggs as more healthful than white eggs—extrapolating from brown rice and brown bread. The color of an egg, however, is determined merely by a hen's breed—brown eggs are laid by brown hens (Rhode Island Red hens usually) and white eggs by white hens (White Leghorn hens usually). The Robertses' eggs were brown, laid by Rhode Island Red hens, and they were no more healthful than white eggs.

Second, labels like "vegetarian-fed" and "omega-3" on eggs are ubiquitous—found even on the Robertses' battery-cage eggs—but they refer only to feed rations and not to living conditions, stating merely that hens have not been fed slaughter by-products and have been fed a diet including omega-3. Brick gave his hens flax seeds for omega-3. Other farms often feed fish, which is an unhealthful practice because fish are not a natural food for hens, and they also leave a smelly aftertaste in eggs.

"Vegetarian-fed" and "omega-3" labels are meaningful only when paired with one of three other labels: "free-run," which means hens are able to walk indoors; "free-range," which means hens have some level of outdoor access; and "organic," which means, among other things, that hens are free-run or free-range. Egg cartons that do not carry one of these three labels mean that hens have almost certainly been confined to cages.

"In the next few years, we're going to go from thirteen thousand hens to eighteen thousand," Paul told me in the hen factory.

The latter number is average for Canadian egg operations and is far lower than the American average of more than 300,000 hens. These numbers are remarkable in and of themselves, but especially when set in comparison to the past. In 1960, approximately two thirds of eggs in North America and Europe came from flocks of fewer than 500 hens.

"Where will you guys put the new cages for new hens?" I asked Paul, looking about myself in dazed confusion. "There's no space for more cages here."

"We won't get *new* cages!" Paul answered, laughing like he'd never thought of such a thing. "We'll just use the same cages."

Oh. This meant that every microwave-sized cage would soon confine six or seven hens instead of the present four or five. It was the human equivalent of a single cot containing six or seven people—all of them eating, excreting, sleeping, and standing in space sufficient for no more than one.

"Do you think hens want to move?" I asked Paul.

"No. Hens are happy in their cages with all their friends."

Science disagrees vehemently. A multinational European government study that examined data from more than 200 egg-laying hen flocks concluded: "Conventional cages do not allow hens to fulfill behaviour priorities, preferences and needs for nesting, perching, foraging and dust bathing in particular. The severe spatial restriction also leads to disuse osteoporosis."

Scientists describe hens as "contra free-loaders"—opposed to "free-loading." A hen's natural instinct is to "work for food rather than accept 'free' food from a feeder." Hens want to forage, in other words, not to fester in a cage.

"These hens have better lives than we do!" Paul informed me. "But when friends of ours from the city come in here and see these cages, they start crying."

Paul laughed like he'd told me a joke. I laughed like I'd heard a joke.

"If my sister came here," he continued, "she'd probably cry, too. I talked to her today. She lives by herself with a pet bird, a lovebird or something. She loves the bird. She says it's her best friend, her 'partner.' She likes it so much that she puts it on the phone when I call her—she makes me talk to it! If I got a bird, it would die on the first day."

What did Paul think he was surrounded by, if not birds? Twinkies? Zombies? His sister had so humanized her bird and he had so objectified his birds that neither remembered any longer that their birds were *birds*—neither people nor potted plants.

On the one hand, Paul didn't seem to view the egg-laying hens as living beings; on the other, he believed that they were happy in their cages. But if they're not living, they cannot be happy, and if they are living, they cannot be happy in cages. There was something psychological at work here: cognitive dissonance—which refers to an uncomfortable feeling caused by holding conflicting ideas simultaneously, and a resulting desire to reduce the conflict by modifying one's beliefs and attitudes.

Standing there with Paul, I was reminded of George Orwell's *1984*, where "doublethink" is the term for believing two contradictory truths at the same time. "The Party intellectual knows in which direction his memories must be altered; he therefore knows that he is playing tricks with reality; but by the exercise of doublethink he also satisfies himself that reality is not violated. The process has to be conscious, or it would not be carried out with sufficient precision, but it also has to be unconscious, or it would bring with it a feeling of falsity and hence of guilt."

Paul seemed to be waiting for comment from me. I kept quiet. The silence became heavy.

"I'm sure that if you got a bird, it wouldn't die on the *first* day!" I finally mustered. It would take a few days.

"Trust me. It would die on the *first* day."

"A motor breaks down just once in five years!" Paul exclaimed.

Located close to the ceiling, on the fourth level of the central column of cages, the motor was responsible for regulating the feed belt for that row of hens. Seeing that it had stopped working,

Paul straightened, cleared his throat, and wiped his glasses. He conferred with Brick in his office, then he hauled out a toolbox and a ladder from the storeroom, all the while whistling. To him, the motor was a blip but also an exciting eclipse in an otherwise mundane universe revolving only around the switchboard, its central star. With the challenge it posed, the motor, like an unpredictable new fling, turned Paul into a new man, a renewed man—a bolt of lightning.

Paul perched on the ladder and lubricated the motor with the black anti-seize paste Brick had recommended. When that didn't work, he unbolted his toolbox and started repairing the motor with his tools. But he was slow and unsteady, for he stopped to cough repeatedly, his face contorting as he barked for breath, his glasses falling askew across his nose.

When I'd first met Paul, I'd assumed he had the flu, the worst case of it I'd ever seen. He didn't, he informed me now. He'd been hacking away just like this for the last thirty-six years. His cough, stemming from both dust and ammonia from hen excrement, had started in his work hours and had, in a matter of years, spread like a tumor to all his hours. He no longer even noticed his cough or bothered to hide it with a hand. It was simply a part of him, like his glasses and his baseball cap.

I'd never felt as nasal or congested as I did every time I entered the hen factory. My lungs clouded with coughs, my nose streamed like an open spigot, my throat itched like it was under attack by a mob of mosquitoes, and my eyes blinked rapidly as they filled and re-filled with dust and ammonia. I'd assumed at first that I'd contracted Paul's flu, until I noticed that my symptoms always started upon entry and sputtered upon exit. I could bear to be in the hen factory for no more than two hours at a stretch, and I couldn't fathom that the hens were trapped there for a full year, inhaling the pungent, poisonous fumes from their own excrement every minute of every day until death.

Before my stay with the Robertses, I hadn't known the word "ammonia" outside of high school chemistry classes. But now every fiber of my being knew the stench well. Ammonia "results from the chemical decomposition of uric acid in droppings," states *World Poultry* magazine. "Partial or complete closure of [hen] eyes are common clinical signs. In severe cases, the eyelids are often closed shut."

"The air of a poultry house can be very dusty," states a book on occupational health and safety hazards, "and for a first-time or occasional visitor, the smell of manure and the pungent odor of ammonia can at times be overwhelming. . . . Acute and chronic respiratory symptoms experienced by poultry workers include cough, wheezing, excessive mucus secretion, shortness of breath and chest pain and tightness."

Paul decided to regale me with a story as he repaired the motor. "This one time, five or six years ago," he recounted, his tone that of narrating a murder mystery, "two *thousand* of our hens died. They all died in the week after we put them in cages. We had no idea why they were dying, so we called the feed company. They sent a guy. He came here and waved his hand in front of them. Hens jump back when you wave a hand in their faces, but they didn't jump back when he waved his. That showed him the problem. The hens were blind!

"It was 'cause of all the ammonia. Not the ammonia here, but the ammonia at the grower. The grower is where the hens grow until they're eighteen to twenty weeks old, which is when we get them, when they're about to start laying eggs. There was so much ammonia at the grower that the hens went blind!"

"But here's the funny thing," Paul continued cheerfully, "nobody at the grower could *tell* they were blind. They'd figured out the food and water system before they became blind, so they kept on eating and drinking like before. They died only when they came here 'cause the cages were a new place for them, and they couldn't figure out where the food and water were."

Though Paul didn't seem to realize it, his terrifying anecdote indicated that hens possessed, at the very least, survival intelligence. They had survived at the grower despite being blind.

"We didn't lose any money on the hens!" Paul exclaimed, misunderstanding my morose expression. "The grower replaced them for free, 'cause the ammonia was their fault."

When Brick saw that Paul's motor-repairing exertions were continuing to prove ineffective, he marched into the hen factory with his younger son, Will. He shouldered Paul aside, scaled the ladder, and hunkered down on the edge of the fourth-floor manure belt. For several minutes, he frowned at and fidgeted with the motor, looking, with legs swinging, like a boy fixing a favorite toy. But suddenly, he froze. There were six hens far down the aisle, *walking*.

They'd probably slipped out of the fingers of the caging crew a week ago, or else they'd scrammed when Paul had opened their cage doors to extract dead cagemates. Looking at the hens, Brick felt that they were taunting him with their freedom. "Can ya and yer girlfriend catch the buggers?" Brick yelled down at Will.

The word "girlfriend" referred to me; to Brick, indiscriminate use of "girlfriend" and "boyfriend" constituted the height of humor.

"Right on," Will called back.

Will strode in the direction of the fugitive hens. I stayed behind. Will turned around and asked me to follow him. I did, unhappily. Will and I walked toward the hens. They split up, dispersing in different directions, disappearing into neighboring aisles, evading us effortlessly. "We're being outsmarted by the bastards," Will stated in summary.

Will devised a new strategy, a "cornering them in the cooler" plan, which was two-part and also two-player, my role in it being scarecrow. I was planted in an aisle to one side of Will so that when he pursued a hen and she ducked into another aisle, she always selected the aisle without me. With Will at her back and me to her side, triangling her, she bolted to the aisle abutting the wall, the aisle with the doorway to the egg-crating room. She proceeded to dash into that room, presuming it a refuge—though it was a ruse.

Coughing and sneezing, panting and perspiring, Will and I eventually succeeded in impelling all the half-dozen hens into the egg-crating room. Will then opened the door to the room's cooler, saying, "Chickens always go into the cooler."

They did, strutting in right away, one after the other, like tourists eager to explore new lands. Will lurched into the cooler and captured them. Clutching them three to a hand, upside-down by their legs, like bags of groceries, he brought them back to the cages. He hurled them in roughly, flinging them like they had no more sensation than rocks. When just one hen remained in his hands, hanging limply, he yanked open the door of a cage—but stopped in his tracks.

Slamming the door shut, he turned around to face me. He walked back to me and held out the hen. "Since you helped me catch them," he said, smiling magnanimously, "you can cage this one."

Instantly, my world changed.

During the hen hunt, I'd acted as a useless accomplice, a caddy to the crime, but if I took the hen, I would change from an abettor to an abuser. I felt splintered in two.

One part of me ordered "*Don't* cage the hen," for confining the hen to a cage would, in effect, be equivalent to twisting or tearing off her legs—she would never walk again. The other part of me countered "*Cage* the hen," for a rejection of Will's effusive offer would be a show of compassion for the hens and thus a slap to Brick's face.

Brick had finished with the motor and was standing shoulder to shoulder with Paul at the foot of the ladder. The two men stared at me with the searing intensity of a laboratory microscope, though none of this was a science, all of it was a sport—a game called caging, the indoor equivalent of hunting. This was the moment of truth, Brick knew as well as I did. I could not fall apart now. But how could I not?

"What's going on?" Will asked me. "What are you waiting for? It won't bite, if that's what you're afraid of."

"I'm just trying to decide which cage to put the hen in," I said, snapping out of my stupor.

My hand reached out rigidly toward Will's. My fingers closed around the hen's legs—tight as a vise, hook-like. So far, in all the time that I'd spent with the Robertses, I'd been surrounded by hens, but I hadn't *touched* any of them. *Nobody* ever touched them. The only point of human interaction with hens was with dead ones, never living ones, when Paul extracted carcasses from cages.

With the hen in my hand, suddenly, all the distance that I'd subconsciously started to cultivate shattered like glass. The hen felt like an extension of my own hand, my own flesh. We were bonded in that moment of contact, chained together in a monstrous web of mutuality. My mental destiny would reflect the hen's physical destiny. Stuffing her into a cage would crush my own sense of compassion and pulverize a part of my own soul. I, and I alone, would be culpable for the condition of the hen. The blame would fall ponderously, punishingly, eternally, on my own two shoulders. A centipede of sweat crawled across my neck.

The hen hung still and silent in my cold, clammy hand, except for strange spasms. The rest of her life would drain away from her day by day in a cage, and so I did not want to confine her to a cage that was already crowded with five hens. I also did not want to confine her to a cage that belonged to the first level or the fourth level. First-floor cages were caves and fourth-floor cages were cemeteries, the difference between them darkness and decomposition. I selected a third-floor cage of four hens.

Unwilling to fling my hen—*my* hen—into a cage by her legs, I turned her upright—*"DON'T!!"* Brick hollered.

But it was too late.

The hen leaned forward, her red-combed head aligning with her back. Her gaze clinched to a point in the distance, and she emitted an angry, ugly cry. Wrestling against my hands, which had unconsciously tightened around her, she unfurled her brown-gold phoenix wings. Before this moment, I hadn't even noticed the hens' wings, for they remained pinched to their sides in their cages. Now, I saw that a hen's wingspan measured *more* than the width of the entire cage she shared with others.

The hen in my hands propelled herself into the air. She flew down the aisle, beautiful to behold. She landed far away and dashed off, disappearing as unreasonably as if she were a magician's trick.

Brick launched into a creative string of curses, incensed at the sight of a tiny hen victory in his tremendous factory. I acted apologetic, but felt ecstatic, for the hen was free (at least for now), and Brick still trusted me.

From his perspective, the hen hunt was a sort of character test, an initiation into a tribe. It was as if we'd gone hunting together. By opening the cage door, I had, in essence, pulled the trigger. I'd failed to strike my target—I hadn't caged the hen—but the outcome was a result not of different character, but different competence. To Brick's mind, there was no longer any distance between us. The hen hunt was a handshake of friendship.

CHAPTER THREE

PORKMAKING BY MACHINERY

FOAM AND FIRE

I had a bitch of a day," Brick Roberts complained. "I was in meetings with Maple Leaf [Foods] academics all day. I hate academics. I can't stand their incompetence. If an academic was starving, I might feed him, but if he was being beat up on the street, I wouldn't help him—not until he was beat up real good—"

"Who are 'academics'?" I interrupted Brick to ask.

"Academics are the guys at companies and universities who sit in their suits in their offices and tell farmers what to do. They told us the high-volume, low-margin model—producing a lot of food, and selling it for low prices—would make us rich. It didn't. That kind of bullshit only makes a few smart guys rich. Like Maple Leaf. And Walmart. Walmart got rich big. Walmart are fuckers. That's why I never go to Walmart. No one should go to Walmart. . . . In our country, food costs less than 10 percent of income, and that's *including* restaurants. In some countries, food costs *50* percent of income. People here buy the cheapest food they can, then they go out and buy three cars and three cell phones with their money. People have lost all common sense."

Brick was right: people spend too little on food, even as they spend too much on all else.

"Do you think of yourself as a family farmer or a factory farmer?" I asked Brick.

"Hmm . . . I don't know. . . ."

Brick's egg operation was a family farm—it was supervised by him and operated by his best friend, Paul—but it was also a factory farm—it looked, from outside and inside both, like a factory. Brick was therefore a *family factory* farmer, a term that is not an oxymoron but a norm.

"All I know is that big farms scare me," Brick said. "Things get out of control and you can't manage them all. But with all the high-volume, low-margin bullshit, farms just keep on getting bigger and bigger."

Every time we spoke, I was struck by how extraordinarily mismatched Brick's profession was from his person. His egg operation with its cages was brutal, but he was himself warm and grumpy, eccentric and endearing. My brown complexion made me an anomaly in his village—I had yet to sight a single soul who was not white—but Brick welcomed me into his fold like a visiting family member, like a niece from out of town. "Ya can call me 'Uncle Brick,' ya know," he said. "Yer a part of the family now."

Brick's children's friends called him "Uncle Brick," and his children called Paul "Uncle Paul." And yet I could call neither man "uncle," despite their hints that I do so. I would try to force myself, but my tongue would trip over the word. One small part of me, I came to see, simply could not reconcile the fact that Brick was a saint to me and a tyrant to his animals. The word "uncle" became the external symbol of my internal struggle to extend my relationship with the Robertses to the level of family.

Brick insisted that I set fire to his garbage.

"There's too much junk in the world to fit into landfills," he explained, "so I light up my junk." He folded into my palm the lighter he kept perpetually in the pocket of his perpetual denim shirt, urging "Make sure ya give it back," his tone suggesting a faint fear that I might run off with it. I obliterated two wheelbarrows' worth of his cardboard cartons to a metallic gray soot.

Multi-faceted Brick was, strangely enough, not just a Walmart-hater but also an environmentalist in his own way, and his concerns about overconsumption and trash pollution affected me. Soon after my stay with him, I read books about these issues. I started composting and recycling diligently, and buying fewer things and using them with greater respect. I began treading on the world more gently, in other words, toward having a lighter environmental footprint.

Another of Brick's passions was his snowmobile. He insisted that I must ride it. I clutched Nick tightly as he zoomed across his family's crystal-white fields like a racecar driver, refusing to decelerate despite my pleas. Rather, the more I wrapped my arms around him, the more he felt spurred to race, until I clung to him more frantically still, and we sat bound together like cemented stones. When the ride eventually ended, we made our way back to our usual hangout—his father's office—to pass the time.

I had just dizzily sat down across from Brick's desk when another activity was recommended to me: curling. I didn't know what it entailed, but I had a feeling I wouldn't like it. "Can we go to a farm instead?" I implored Nick. The only two farms I'd seen so far—Michael Miller's dairy and Brick's egg-laying hen factory—were very different from one another, and I was curious to see what a third sort of farm might look like.

Nick looked to his father, his glance asking "Can we trust her?"

Brick, who was cursing at his printer, nodded absent-mindedly.

Nick placed a call on his cell phone. "Hey, Charlie, how's it going? . . . Right on. We have a friend visiting, and she's interested in food production. . . . Why? I have no idea. You can ask her yourself. . . . Don't worry—we trust her. . . . Can we—me, her, and Will—come over to your pig farm right now? . . . Cool. We'll bring beer, and, yeah, that other thing, too."

Brick cautioned me about the visit. "I wouldn't go to Charlie's pig farm if I were ya," he said. "His pigs aren't happy. I wouldn't eat them. I'd rather eat a happy pig. Charlie's pig farm will smell real bad. And ya'll have to shower in and shower out because of biosecurity. Biosecurity is bullshit. We're also supposed to make ya shower every time ya see the chickens—both before *and* after ya see them—so ya don't kill them with yer germs. But if we thought ya were going to kill them with yer germs, would we let ya see them? No, we wouldn't! We didn't make ya shower, but they're still alive; that's why biosecurity is bullshit. But don't go telling any of our friends we said that. If Charlie or any of our other friends ask, tell them we *always* made ya shower in and shower out."

Nick's truck was littered with depleted beer bottles, soda cans, and magazines and papers, including a cut-out, taped to the glove compartment, of a woman in wispy panties. Nick and Will settled into the driver and passenger seats, each with a bottle of beer in one hand and a cigarette in the other.

"Thank you for being my farm tour guides," I said.

"I'm sure we'll graduate to friends soon," Nick replied with a chuckle. As a forewarning of his friendship, he stared at me in the rearview mirror throughout the hour-long drive.

We reached Charlie's house after sunset and were greeted by an elderly Australian Shepherd, who barked quiveringly at the approach of the odd squad. Standing beside her was Charlie, lanky as a lamppost. Twenty-five years old, he had a donut-round face frosted with sandy curls and blazoned with a boyish grin. He invited us in for beer.

Charlie's living room was cozy but mushy, its walls and shelves adorned with dewy-eyed pictures of him and his wife holding hands and locking gazes. Charlie showed me his wedding photo album page by page, kept on the coffee table for easy reach. He was friendly and jolly, oozing with romantic sentiments that appeared more suited to a poet than a pig farmer, and I found it impossible to reconcile the cheerful man before me with Brick's scathing portrayal of his farm.

After the three men had chugged a few beers and Will had taken a "whiz" on the front wall of Charlie's house, we piled into Charlie's truck for the drive to his farm. Then Nick reached into his shirt pocket and extracted a joint of marijuana.

"Don't tell Brick we smoke dope," he told me with a laugh. "Brick got mad when they showed a hillbilly farmer smoking dope in *Harold and Kumar*. But we hillbilly farmers love dope!"

From the outside, Charlie's pig facility looked even less like a farm than Brick's hen factory. It looked like a series of interconnected warehouses.

As Brick had warned, Charlie directed us to shower as soon as we entered. But the order was superfluous, for there were only two paths to proceed farther from the entrance: the men's and the women's shower stall. A sow (female pig) smiled brightly from the cartoon picture taped to the women's shower door, and a boar (male pig) reclining in a tub of mud grinned from the men's door.

"Do you mind if I don't wash my hair?" I asked Charlie. "I don't want it to be wet in the cold."

"You *have* to wash your hair," Charlie said. "It's part of biosecurity protocol. Wear the clothes on the other side of the shower, and the underwear, too."

I did rinse my long hair, but I didn't shampoo it, in part to protest the intrusive protocol and in part because the blue grid mat on the shower floor cut into my bare feet. After my shower, I wore a pair of loose, dark blue women's scrubs; underneath them, feeling rebellious, I retained my own underwear instead of donning public panties. The Roberts brothers also defied their friend's underwear decree, but more sensually. They omitted undergarments, including undershirts, altogether, the plunging necklines of their matching gray overalls showcasing rolling prairies of chest hair.

By the end of the biosecurity process, Nick, Will, and I appeared attired for incarceration. In addition to shedding all layers of our clothes, we'd also changed our shoes, substituting for them with knee-high black boots. Giving us a quick look-over, warden Charlie opened the next door.

The corridor before us was long and labyrinthine, gray and cold, with peeling plaster walls. It looked like it would lead us to a dungeon.

It did.

Like school classroom doors, each of the several doors in the corridor contained a window, but the windows could just as well have been shrouded in an opaque black cloth, for, devoid of light, the rooms were impenetrable by sight. Eventually, Charlie stopped at a door and opened it, stating "This is a farrowing room."

Because the farrowing room was black, I could not see anything, but I could smell. "Charlie's farm will smell real bad," Brick had warned, but this was an immense understatement. A poisonous potion of ammonia, hydrogen sulfide, methane, and other stinking manure gases swirled around us. It was the feeling of fingers of feces poking my nostrils, of a fist unloading feces into my mouth. I felt like I was drowning in dung, like the pressure cooker of manure gases—brimming, bubbling, bristling—would swallow me whole, turning me into a mound of excrement myself. The smog of stench was so heavy and pressing that, feeling instantly sick and claustrophobic, I decided to breathe only from my mouth for the length of the visit.

When Charlie found the light switch, low-hanging yellow lamps flickered on, illuminating under them sixteen sows. The sows were gargantuan creatures, weighing 400 to 500 pounds, measuring from snout to tail nearly six feet in length. Each was confined to a narrow metal contraption called a crate. The sows lay on their sides in their crates, their flesh bursting out from between metal bars, their legs parked outside their crates, for the crates disregarded the animal instinct to be *animate* not just conceptually but also practically: they left not an inch of space for legs. The sows were just torsos, supersized slugs. They could not walk a step.

Each rectangular crate sat like the center of a sandwich in the middle of a rectangular enclosure. Packs of piglets, clustered together like magnets, nursed or slept or stirred in the sides of the enclosure. Sows birthed throngs of piglets—about a dozen at a time—but they reared none of them, for the bars of their crates prevented them from licking and grooming their young.

The crates could also be called coffins. Some sows looked dead—they didn't even seem to be breathing. I tremulously asked Charlie whether they were alive. "I'll *show* you that they're alive," he promised with a laugh.

Charlie pulled on to his right arm a glove that stretched all the way to his shoulder, like the glove worn by Arthur, the artificial inseminator at the Miller dairy farm. He crouched onto his knees behind a laboring sow and forced his arm into her womb—past his elbow. The sow, who'd appeared dead, jolted to life like a defibrillated patient. She jerked, twisted, screeched, struggling to escape Charlie's arm, but unable to, for her crate trapped her firmly in place.

"I'm trying to take piglets out," Charlie told me. His aim was to speed up the birthing process, but his hand returned with only blood every time, of this sow and all the next laboring ones he assailed. Will Roberts, in the meantime, helped Charlie by injecting sows in labor with oxytocin hormone. Smiling and whistling, he appeared to enjoy the task of sticking a needle anywhere in their bodies.

Charlie eventually tossed aside his bloodied glove. I hoped this marked the end of the rounds, but it didn't. He started injecting sows with antibiotics. Directly above every sow's head sat a clipboard depicting a full-page grid graph scratched with X marks in blue. Each X signified the sow's food intake on each of the twenty-one days that Charlie permitted her piglets to have access to her teats, before he forcibly weaned them.

He picked up a clipboard and studied it. "When the sows don't eat enough," he informed me, a twinkle in his eyes, "they get a needle of penicillin."

Returning the clipboard to its place, Charlie injected the corresponding sow with penicillin. Like Will with the oxytocin injections, he did it carelessly, clumsily, without seeking a specific site of her body, his penicillin administration like flinging a penny into a fountain. The sows—viewed as fountains of piglets—remained as leaden and insensible as before, no longer even noticing their daily needles, having become, as was intended by the experiment of factory farming, less "live" and more "stock." This sow facility, and the countless others like it in North America, are laboratories gone horribly wrong, turning men into monsters and living animals into half-dead objects. Charlie and Will looked like mad scientists.

By shooting sows full of penicillin—the first antibiotic ever invented, the most iconic antibiotic, the vital, embattled grandfather of antibiotics—Charlie contributed to antibiotic resistance. Worse, he did so not to duel disease, but to instill appetite: antibiotics make farm animals gain weight faster. But sub-therapeutic doses of antibiotics are dangerous because they provide enough medicine to kill some but not all bacteria, and the remaining bacteria become more resistant, stronger, posing a greater danger to both animal and human health. Use of antibiotics in farm animals accounts for more than 70 percent of all antibiotic use in the United States. Essentially, human lives are put at stake so the agriculture industry can generate higher revenues for itself.

Charlie's breeding pigs, in contradiction with their nature—in contradiction with their very name—had no desire to "hog" or "pig out." That one fact about the sows—their lack of appetite, their failure to "show up for a meal"—would have led E. B. White, pastoral pig farmer and author of *Charlotte's Web*, to diagnose the sows as ill, especially in light of their having no "showing up" to do.

Ground corn and soy, mixed in water for ease of transport through overhead pipes, showed up before their snouts twice daily, dumped with a splash onto the red trays affixed to the front of their crates. But the sows, moored like broken boats above an underground pond of festering feces, were indifferent to the frothing broth, acting more extinct than extant even at mealtimes, apathetic to everything except Charlie's insistent forearm in their rears.

Though the sows were medicated excessively, their piglets were not medicated even when necessary.

Charlie reached a hand into a litter and plucked up a piglet, dangling him at my eye level upside-down by his hind legs.

His eyes terrified, the piglet squirmed like a fish on a hook, and he shrieked, the sounds panicked and piercingly high-pitched. He'd been mutilated by Charlie earlier in the day. Charlie had sliced off his tail, leaving a stub the length of a fingernail, and he'd castrated him, leaving a stinging, raw, ruby-red scrotum.

"I castrate them by cutting their scrotum open with a knife," Charlie detailed, "then I put my hands in and pull out the testicles."

"Do you use anesthesia?" I asked.

"No. Castration doesn't hurt at all."

But it hurts enough that Europe has created a *European Declaration on Alternatives to Surgical Castration of Pigs*. "It has been scientifically proven, using physiological and ethological parameters," states the Declaration, "that surgical castration is a painful intervention even when performed on very young animals." The Declaration instructs that pig castration should be conducted with anesthesia, when conducted at all, and should be abandoned entirely by the year 2018.

Unexpectedly, Charlie passed the piglet to me. He screeched frightfully and wriggled frantically, as though there were a noose around his neck. I returned him to his enclosure as fast as if my hands were on fire. Charlie and Will sniggered at my queasiness. The piglet, in the meantime, bolted to his mother's belly. But she was a ghost mother, a specter; she'd slept throughout his alarmed episode.

We left that first farrowing room, and entered the corridor again. It was lined with pails filled to the brim with newborn casualties of castration, stillbirths, and other deaths. In the pails, the hundreds of dead piglets looked like globs of pink Jell-O floating in a creamy custard of blood. They were a macabre sight—having died before they'd ever lived—and I couldn't bear to look at them.

For the next hour, we entered some more farrowing rooms. All of them matched the first in setup and stink, and Charlie proceeded in all of them with his painful sow midwifing and medicating. Then we entered a different kind of room, which Charlie called a "gestation" room.

The farrowing rooms had each confined sixteen sows; the gestation room confined five hundred. It was gigantic.

The sows in the farrowing rooms had been giving birth; those in the gestation room were pregnant, or to be made pregnant. When the lights went on, the gestating sows rose to life by biting the bars of their crates, smashing their heads against them, and screaming. The racket was as deafening as walking into a room with multiple fire alarms ringing simultaneously. I covered my ears with my hands in the bedlam, biting my lip to suspend my own scream, which threatened to mingle with that of the sows.

The sows' chaotic cacophony was startling in its intensity, but even more so in its quality. The hysterical, high-pitched howls sounded eerily human—like "the shriek of a man," in Leo Tolstoy's words. But a lunatic man, for the sows—splattered with graffiti-like blue, green, and pink letters and numbers—seemed insane in their stationary rampage.

Charlie began walking down an aisle, and Nick, Will, and I followed him single file. The sows all around us existed in four long rows of gestation crates. As narrow as my narrow shoulders, the crates were equivalent, in Dr. Temple Grandin's words, to forcing "a sow to live in an airline seat."

Like the farrowing crates of the previous rooms, gestation crates allotted not an inch of space for legs. When a sow lay on her side, her legs lodged entirely in her neighbor's crate—cramping her neighbor, hampering her, poking into her back or her belly—as she did the same with her neighbor to her other side. Nine tenths of the millions of sows in the United States and Canada spend their lives confined to such crates.

Charlie's sows were boxed in from all four sides, imprisoned by bulky beams not just to their right and left, but also above. I asked Charlie what the overhead beams were for. He couldn't hear me above the clamor. I tapped his shoulder to halt his stride, then hollered into his ear.

"If you give a sow six inches on top," he shouted back, "she'll try to jump out of her crate. She'll step on her feed trough to try to jump out. She'll never give up. That's why she has to be closed in completely. The crates have to be really strong to keep her in. They're made of iron-galvanized steel. The steel's good and strong, but the crates are dipped in iron, too, so they don't rust as fast in all the ammonia. Without the iron, they'd rust in a year, tops. There's a *lot* of ammonia in here."

"Why do you keep sows in crates instead of housing them together?" I yelled back.

"If you loose-house them, they can fight."

The logic, in other words, was to restrict *all* behavior in order to suppress *one* behavior. Would it not be more sensible to better a sow's social disposition by improving her living condition?

The sow protests, in their vehemence and violence, in their might and misery, were reminiscent of *The Jungle*. "The uproar was appalling, perilous to the eardrums; one feared there was too much sound for the room to hold—that the walls must give way or the ceiling crack," wrote Upton Sinclair of a pig slaughter plant more than a century ago. "There were high squeals and low squeals, grunts, and wails of agony; there would come a momentary lull, and then a fresh outburst, louder than ever, surging up to a deafening climax. . . . It was all so very businesslike that one watched it fascinated. It was porkmaking by machinery, porkmaking by applied mathematics. And yet somehow the most matter-of-fact person could not help thinking of the hogs, they were so innocent, they came so very trustingly; and they were so very human in their protests. . . . It was like some horrible crime committed in a dungeon, all unseen and unheeded, buried out of sight and of memory."

Perhaps it was my wrought emotional state; or perhaps it was the claustrophobic feeling of being trapped in a dungeon; or perhaps it was the fetid, fecal stench of ammonia, devastating my mental mettle faster than crate metal; or perhaps it was dawning comprehension of why Brick Roberts, who caged hens, pitied pigs; or perhaps it was newfound understanding of why Jane Roberts had developed a lifelong aversion to eating pigs when she'd worked at a sow facility; or perhaps it was light-headed hunger—I'd been famished during the drive to Charlie's breeding facility, but now I could not imagine ever eating again; or perhaps it was simply the observation that the sows were so numerous, and so noisy, and so neglected, and so *near*.

For whatever reason—in bold disobedience of Charlie, who warned "The sows *don't* like being 'petted'"—I suddenly wanted to stroke a sow. My hand slipped through iron-galvanized steel bars to descend on a slender, scraggly pink back coated with green writing and black flies.

The sow yelped and leaped as if electrocuted. I yelped and leaped. Nick rushed to my side.

"*Don't* do that again!" Charlie berated, swatting flies away from his face. "As I said, the sows have never been '*petted*,' so they *don't* like it. If this sow had jumped *one* inch higher, your hand would be broken right now—it would be crushed between her back and one of these overhead bars."

The sows were like the knives in a knife block: crowded but isolated, surrounded but separated, unable to interact with people or with pigs. They couldn't even mate.

Like Michael Miller's dairy cows, Charlie's sows were artificially inseminated. That said, in a crate at the very front of the gestation room lived an immense, fierce bear of a boar. He played two essential roles in the piglet production plant, Charlie explained, both of them too chemical and complex to be replaced by man or machine.

He was necessary, first, because a sow's reaction to him told Charlie whether or not she was in heat, and, second, because he "turned sows on," which latter was a significant difference between the insemination of a sow and a cow. A cow could be impregnated without being interested or aroused; a sow couldn't, not if she was to produce a large litter. The boar was retained for arousal purposes.

He was transferred daily from his enclosure to a crate with wheels and paraded through the aisles of sows like a rock star on tour. A sow who was not in heat looked straight through him, as through a window. A sow in heat ogled him lustfully. Her ears popped up and she sprang to her feet to stand still as a stick for him. Her eager, inert position was not called "standing for the boar," however, but "standing for the man," for it was not the boar, but a man—Charlie or a worker—who crept into position behind her.

The man would lower onto her rump a C-shaped metal contraption with yellow tips that would squeeze her heaving sides to imitate the pressure of a mounting boar. The man would then thrust into her a semen catheter, and he would consummate the act by recording on the clipboard above her the date of artificial insemination and the expected date of parturition. The boar, in the meantime, would be rolled along down the aisle of sows.

"Poor guy!" Will said. "He has these big *cojones*, and he's around all these females, but he can't *do* anything about it."

Nick nodded in grave-faced agreement with his younger brother.

Charlie hurriedly assuaged their unease by acknowledging that, in his own sense of affinity with the big boar, he did occasionally let the boar "*do* something about it." He let a sow into his crate sometimes.

Nick and Will breathed a sigh of relief.

We left that first gestation room and then toured another, a twin to the first except for more flies. Finally, we showered. I'd never been so eager to scrub myself clean. Showering prior to entry had felt absurd, like getting a manicure before plunging my hands into the toilet, but prior to exit, the shower felt vital. The relief of washing was diminished by discomfort, however, for the blue grid mat on the shower floor cut painfully into my feet. It was like the metal grill above a street gutter, all but stamping its design on the soles of my feet.

As I glared at the blue mat, I realized that that same mat also served as the floor of farrowing crates. Throughout the four weeks that sows were confined to farrowing crates, they stood and slept on that mat. It hurt their legs just as much as it did mine. Many of them became injured and lame as a result.

The life of a present-day sow is a cycle of protracted misery. To impregnate the sow sooner—to drive her faster, to benefit from her better—pig farmers wean her piglets prematurely, at three weeks of age or earlier. Then, just a few days after weaning, they penetrate her with a catheter of semen. Fast-forward almost four months, and the sow—the piglet-producing machine, one of countless cogs in the giant pork production complex—is ready to give birth. She stumbles on obsolete, crippled legs through the corridor from her gestation crate to a farrowing crate. There, lying on a metal grid mat, injected daily with antibiotics, she gives birth to a litter of piglets, and the cycle commences anew.

In addition to his sow operation, Charlie owned also a "finishing" one, where piglets were "finished" for slaughter, which would be at six months of age.

Charlie's finishing facility was our next destination for the night, reached after a brief but busy truck ride involving beer, pot, and cigarettes. The three men basked in one another's company, making jokes about the Mennonites we passed on the road, poking fun at their horse-drawn wagons and ankle-length dresses.

When we reached Charlie's finishing facility, he did not order us to shower again, instead noting "You've already showered twice tonight." He did, however, instruct us to change completely out of our clothes. I didn't. Garbed in goose bumps, I shrugged farm scrubs on over my own sweater, hoping Charlie wouldn't notice.

He didn't. In the farrowing facility, he'd been stern, but he was now smiling and soft-toned: *stoned*. He, Nick, and Will carried into the finishing facility beer and cigarettes, even though all three men knew that cigarettes in pig facilities pose extreme fire hazards.

Nick and Will Roberts knew this from family experience. Brick had, at their age, been a sow farmer, and he would have continued to be one today had not a fire—one sleepy night before Nick and Will were born—swallowed his sows and their litters of little ones. Screaming and writhing, the sows had sweltered and smoldered to death.

Today, pig farms have been catching fire everywhere, from Iowa to Minnesota, from Ontario to Quebec. The catalyst of conflagration has been determined to be a slimy, gelatinous layer called a "foam," which forms on manure in underground pits. Wobbling and throbbing like it's alive, the foam functions by trapping evaporating fecal gases and holding them hostage until it's agitated. Then it explodes like a bomb, releasing the gases so fast and so furiously that they burst into flame as they flee from underground pits, burning on their way upward the desperate, trapped creatures who birthed them.

According to surveys, around a quarter of pig operations in the agricultural-intensive states of Iowa, Illinois, and Minnesota have foam, putting them at serious risk of a fire.

As the men's cigarettes glimmered gold—innocent but ominous, shimmering but menacing—my imagination kindled with visions of being swallowed, like Brick's sows, by a storm of smoke, lashing and thrashing in vain, never to emerge again.

I forgot all about foam and fire, however, when we stepped into the passageway, and Will tried to electrocute me. At Charlie's finishing facility, the corridor was as dungeon-like as that of the breeding building, but it was made more forbidding still by electric prods. Two feet long, they posed like lampposts against the walls, prepared to ignite all pedestrians, humans and hogs. As Charlie described to us that his finishing facility consisted of four identical rooms, sorted by age, each room divided into forty-four pens, Will yawned and pawed an electric prod.

He probed it, twirled it, savored its constrained power in the palm of his hand. He pointed the prod at me and pushed the button. A lightning flash cracked the black—white-blue, blue-white.

I leapt back, shocked mentally if not physically. "What's *wrong* with you?" I brayed.

Will's lips parted into a silly grin.

When I'd touched the sow earlier in the evening, Charlie had pounced on me like a cat on a mouse, but he didn't even frown at Will's galvanic whims. "The electric shock really hurts," he drawled dreamily, mellow as a kitten. "I've gotten myself a few times by accident when moving the pigs. Prods make them move faster, but you don't want to get shocked by them yourself."

The sow I'd touched had reacted precisely like she'd been electrocuted. She *had* been electrocuted, Charlie's words suggested, probably on trips from the farrowing room to the gestation room, and then back from the gestation room to the farrowing room. Her apparent assumption that any touch to her back would be charged suggested that she'd been shocked repeatedly in her life.

Charlie opened the first of four doors.

The room was an enormous rectangle, sundered into two sections by a wide aisle down the center. There were twenty-two pens to each side of the aisle, and twenty-one pigs to a pen, totaling close to a thousand pigs in the room.

The pigs, two to three months old—large piglets, really—seemed to have been playing. They arrested all activity upon our entry, like school children upon the entrance of a teacher. They pierced us with stares, many of them trotting up to the hip-high concrete walls of their pens to stare better at the strangers. This room was precisely as barren as the sow farrowing and gestation rooms, the air tossed and toiled just as laboriously with ammonia, but the pigs acted neither dead nor rabid like their mothers. To my relief, they were *animated.*

The pigs in the second room were noticeably less animated. At three to four months of age, they were a month older than the pigs in the first room, and they were bigger—the length of a pillow instead of a cushion. They lived, however, in pens of the same size. They were consequently more cramped and squalid, their skin a pink sky shrouded with clumped, crusted clouds of feces.

In the third room, the pigs were a month older still, and their circumstances were exponentially worse. From head to hoof, they lived marinated in manure. At four to five months of age, they weighed as much as adult men—up to 200 pounds—but they were accorded fewer square feet of space each than human toddlers in cribs. The brown floor slats underneath the pigs lay submerged under feces and flesh, and could be spied only when, after a gushing, whooshing, waterfall-like sound overhead, the narrow troughs in the pens filled with corn-water, and the pigs jostled to the troughs, blaring and broadcasting their passage like police-car sirens.

The final room would have been even more thronged than the third had nine tenths of its residents not just been trucked to slaughter. Their killing was their final punishment but also their first present, their final chastisement but also their first charity, an end both early and overdue. They were taken late at night so that, half-asleep, they could be more easily pushed and electrically prodded onto three-story trailer trucks that crammed together 240 pigs. Calmer and drowsier though the pigs may have been at night, the timing of their journey meant that their brief existence had ended without ever having sighted the giver of life, the bosom that sustains the earth—the sun.

The pigs' slaughter created unexpected benefactors, however. The pigs remaining in the room lived now no more than five to a pen, and they advanced without accident—galloping, galumphing, triumphing. There was still room in their enclosures even for us, and Charlie urged us to climb with him into a pen of five pigs. I climbed in hesitantly—but regretted it when I was charged from behind.

My upper thighs seemed suddenly trapped in the jaws of something strong and forceful. As I turned around with a gasp,

I saw that my assailant was not who I'd expected it to be. It was not a grunting pig, but a leering, smoking Will. Drunk and high, Will had noticed my expression of consternation—I was uneasy in the pen of pigs, recalling the panicked sow I'd touched—and he'd decided to stoke the unease by grabbing my legs in imitation of a charging pig.

I didn't much like Will, I decided.

I reddened. Charlie snickered. Nick glared at Will.

The resident pigs stared, from their position at the back of the pen, where they'd fled as a unit when we'd climbed in. They peered at us for a few minutes then scuttled over together, a grunting swine squad. They sniffed our boots, then chewed them, resisting our toe taps to dislodge them.

Better they chew boots than backsides, Charlie commented. When denied any and all diversion, as they were, the pigs developed a terrifying tendency, he explained. They turned the tails of other pigs into toys, biting them the more they bled.

Charlie snipped their tails from ringlet curls to stubs as soon as they were born, but the reason for the mutilation was counter-intuitive. Tails were docked not so that there would be less to bite, but so that the remaining stub would be more sensitive, and pigs would be more motivated to avoid being bitten. Otherwise, if the tail were full and healthy, pigs would let it get chewed due to learned helplessness. In other words, tails were docked not to influence the behavior of the biter but of the bitten.

A better solution than tail-docking would have been to reduce crowding. "Most kinds of aggressive behavior among members of the same species are responsive to crowding in the environment," states naturalist E. O. Wilson in *On Human Nature*.

As Charlie stood *with* the pigs—human to hog, thigh to shoulder, in their world, their enclosure—he seemed to transform under their slobbers from their god to their guardian, their castrator to their custodian. Reaching down, he patted the head of a pig at his boot.

"Pigs are smart," he said, his tone now tender, reverent. "They're really curious, too. They like smelling new things, touching new things. They learn to share, they figure things out. They're also really clean. If you look around, you'll see that all the pigs in this room have made the back of their pen the bathroom. The pigs in the other rooms would do that, too, if they had the space."

In George Orwell's political allegory *Animal Farm*, it was pigs who were commanders of the farm, since they were "generally recognized as being the cleverest of the animals."

According to an article in *Modern Farmer*, an agricultural publication, pigs are social, have a good memory, remember experiences, can tell the difference between individual pigs and humans, and can learn from one another.

Pigs are among the most intelligent of animals, but among the worst treated of them.

In Charlie's truck, a blend of cigarettes, marijuana, and ammonia created an invisible but palpable cloud, aggravating the already trumpet-loud coughs we'd acquired over the course of the evening. Every movement of my head, hair, and sweater made my nose wrinkle with my own stink. Charlie, Nick, and Will had changed completely prior to entering the finishing facility not just because of biosecurity, I realized now, but also because they'd predicted the imprint of ammonia. I hadn't, and was now paying for it.

"Ammonia doesn't go away," Nick said. "Even after pig farmers shower, you can still smell the ammonia on them if you get close. And you *really* smell it when they sweat, 'cause it's gotten all the way into their pores, and it comes out when they sweat. You don't want to be near a sweating pig farmer!"

But one would learn none of this from the website of Conestoga Meat Packers, the "vertically integrated processor" that Charlie supplied, along with more than 150 other Ontario pig growers. The pigs we'd just seen would be trucked to Conestoga's slaughter plant, where they would be killed and converted to hams, bacons, and sausages—which would be sold in more than thirty countries, including Canada, the U.S., Japan, and Korea. Consumers would eat pigs in blissful ignorance of their stinking, festering conditions.

Conestoga's marketing name is different from its official name, the logo on its meats reading "Conestoga Farm Fresh Pork." The pictures on Conestoga's website complement the "farm fresh" in its logo. There are no steel crates, no concrete pens, no electric prods. There are no pigs either. Not *one* picture depicts *one* pig, a curious fact for a pig processor.

Though Conestoga's images balk at illuminating the production methods behind "the world's best pork," Conestoga's president, Arnold Drung, does not. "The key to efficiency," said Drung in an article titled "Efficiency rules at Conestoga Meat Packers," "is getting maximum yield from our inputs."

"Maximum yield" is obtained through "complete control" of everything: hair and underwear, showers and shoes, feces and food, penicillin and oxytocin, castration and insemination. Conestoga even has a video to sermonize about the necessity of "maintaining vigil on the control philosophy."

From the window of Charlie's truck, I spotted the silhouettes of some animals up ahead. They were too tall to be pigs, and too slender to be cows, and yet, standing behind a wire fence, they could not possibly be what they seemed. My eyes were seeing wrong, I decided. The night was black, there were no street lights. . . . We neared. "*STOP!!*" I cried.

The truck jolted to a halt. Nick, Will, and Charlie turned to stare at me like I'd sprouted antlers.

Deer. About fifty of them. Behind a wire fence. A *deer* farm.

I didn't know it then, but there are hundreds of deer farms in North America, their ranks expanding every year.

As I looked at the deer, elegant and alert, my mind saturated simultaneously with thoughts of the past, the present, and the future of animal agriculture. Less than half a century ago, pigs had lived like these deer—outdoors, in the open air, under the sun and stars. Today, pigs are confined from birth to death in closed factories, out of sight and out of mind.

If people develop a hankering for venison as they have for pork, will deer farming explode from a wild venture to commercial agriculture? Will deer become the pigs of the future? Will they be bred to grow as big and bloated as pigs, and as briskly? Will they be confined to barrack-like buildings with thousands of others? Will breeding females be immobilized in steel crates? Will they be artificially inseminated, then artificially separated from their fawns?

When will it begin, for deer? And where will it end, for pigs?

CHAPTER FOUR

TRILLING TURKEYS

THE FREE-RANGE FRAUD

With its wall décor of hockey jerseys, Nick Roberts's favorite restaurant was more elegant than The Truck Stop favored by Brick, but it was also more sparse in patrons, for whose absence in ambience it compensated with television. Several flat screens flashed a hockey game.

Nick chewed through two dozen chicken wings, some of the red barbecue sauce sticking to his beard. "It's so messed up, you can't drink when you drive. It's Communist. I always drink when I drive. So what? Why not? . . . When I take over the egg farm, I'm not just gonna keep on doing what Brick's doing. I'm gonna grow the egg farm a lot. I'll get a lot more hens."

More animals. Why did everybody talk about more animals? Charlie the pig farmer had twice as many sows as his father had, and he would keep adding more to his inventory every year. Nick had similar plans with egg-laying hens.

Nick seemed to want to say something, but didn't know how. "I was watching you at Charlie's pig farm," he eventually muttered, his face scrunched into a frown. "You think there's animal cruelty on farms, don't you?"

"No, I don't!" I lied.

Nick breathed a sigh of relief.

"Right on," he said, smiling. "Just thought I'd ask. . . . It's great you're so into learning about food production and our way of life. I'm glad you came to our house. You're a good influence on me. Like, with swearing. I didn't notice I swear all the time until I met you, and saw that you don't. You're the only person I know who doesn't swear. I'm gonna stop, too. And I'm also gonna stop smoking—I can tell you don't like it."

Nick really had been watching me closely.

"So, you know all about me. Tell me about you. What's your type of guy?"

"Who knows?" I gulped.

Nick's face crumpled like an autumn leaf, as if he'd expected the response to be "*You*, Nick! *You*!"

"I've been meaning to tell you something else, too. It's not right for you to take the bus home when you leave. Stay with us as long as you can, then I'll drive you back to the city. We can spend the day together. You can show me around, we can go for a movie, we can go to your place. . . ."

I didn't like where this was going. I valued Nick as a friend, nothing more; we were like water and oil—we didn't mix. "Thanks, but definitely don't worry about dropping me back!"

"It's no problem. I'll enjoy it. I'm already looking forward to it."

When the check arrived, placed at Nick's elbow discreetly by the waitress, he was startled to see me lunge for it. "That's not how a date works!" he protested, scrambling to snatch it out of my hands. My purpose in paying was to squash Nick's idea that we were on a date; it backfired. Nick instead inferred intense interest. Why else would a woman pick up the tab for his two dozen chicken wings?

"Let's go to my place for a drink," he said in his truck.

"I'm tired."

"Okay. Tomorrow night?"

DO NOT ENTER WITHOUT AUTHORIZATION, warned the yellow sign on the metal door. BIOSECURITY IN EFFECT. Nick's mother, Jane Roberts, pushed the door open.

White, tiny, fluffy three-week-old turkeys, 4,500 of them, surged toward us until, in mere moments, they'd formed froths of feathers around our feet. They fanned forward and backward, joining and separating, like waves rushing and receding around rocks. Jane and I waded into the welcoming ocean of feathers, lifting our feet slowly and lowering them lightly to avoid crushing the little turkeys.

"The baby turkeys follow me around," Jane said proudly. "They recognize me."

Snaking across the length of the shed were three metal pipes. On one of them dangled large, red, bell-shaped feeders filled with a granular yellow corn mix. The other two pipes held small green water containers. Sunlight in the shed was scant—one wall was lined with shuttered windows and the other newly fitted with solar panels—but the place was well illuminated by yellow lightbulbs. At 84 degrees Fahrenheit, the air was sauna-warm, heated by two dozen wide, metallic, umbrella-shaped heaters called "brooders."

"When baby turkeys sit under brooders," Jane said, "it's like their mothers are sitting on them. That's why they're called 'brooders.'"

Metal brooders can hardly be replacements for mothers, I found myself thinking.

The turkeys were themselves "little heaters," as Jane described, having a higher body temperature than us. Our feet soon became pleasantly toasty from their trodding. The wood shavings on the floor were clean enough to sit on, and Jane and I settled down under a brooder, stretching our legs before us as though we were at a beach. The young turkeys plopped down around us in huddles, as water pools in puddles. I plunged my hands into the sea of feathers and snatched up two of them, one to a hand. Close-up, they didn't look as healthy as they did from a distance.

They were declawed—their feet, which flailed through the air in protest, were protuberances of pink, more naked and noticeable than human feet without toenails. They were de-beaked—their beak without the tip looked like a pen without a point, like a precipice without a peak, like an egg-laying hen's beak. They were de-snooded—the snood, a red rope of fleshy wattle that would dangle over the beak (to attract females) as they would grow older, had been nipped in the bud.

The turkeys' claws were removed and their beaks blunted because, as they would grow older, life in the turkey shed would become far more crowded, and they would be more inclined to attack one another. The snood was snipped off because that was the favored point of attack. Both the tools and the target of attack were eliminated.

"Aren't the baby turkeys cute?" Jane asked me.

From the start of my residence with the Robertses, terrified at the thought of inciting another antagonistic conversation like the one that had ruined my stay with the dairy-farming Millers, I'd abolished adjectives like "cute" from my dictionary. It had been my compliment "your calves are cute" that Michael had found unbearable. Sitting among the turkeys, I didn't know whether Jane's question was sincere or a test. She often seemed to be trying to gauge my reaction to things, and then trying to decide what to make of that reaction. Brick and Paul trusted me completely, but Jane, I could tell, didn't.

"Yes, I guess the baby turkeys are cute," I concurred hesitantly.

Jane smiled. I got the sense that her question had been sincere.

The Roberts turkeys were free-range. Or that's what their label would say when they would be sold in supermarkets. The batch of turkeys sprawled around me wasn't *actually* free-range, though. Normally, the Roberts turkeys were let out into an attached yard after six weeks of age, but the fence bordering one side of the turkey shed had been flattened by a truck. The turkeys couldn't be let out until after the fence was fixed, and it wouldn't be fixed until after they'd already been slaughtered. This meant that they wouldn't "range" for one day of their lives, but they would still be described—and priced—as free-range.

The free-range farce was startling, but also not startling once I came to know the full free-range story. Brick became a free-range turkey producer after his pig farm burned down.

The pig-farm fire forced twenty-five-year-old Brick to take a sojourn from his four interests—drinking, smoking, gambling, and chasing women—in favor of constructing a new path for the future. Should he build another pig farm, he wondered, or should he treat the fire as an opportunity to enter another area of animal agriculture, perhaps in another way?

Brick's consideration of the question was affected by two realizations, the first that "pigs don't make any money," and the second that "neither do eggs." His pig farm had burned down on its own, but Brick couldn't very well burn down his egg factory. He *could*, however, ensure that his next agricultural endeavor adopted an opposite paradigm to the egg operation—that its economic model was "low volume and high margin" instead of "high volume and low margin." This meant alternative agriculture: organic or free-range.

Organic involves certain minimum stipulations, among them that farm animals eat organic feed, which is more expensive than conventional feed, and that they have 120 days of outdoor access a year. Free-range, in contrast, is not properly defined in the United States and Canada. The term means merely some vague level of outdoor access. Important questions such as how much outdoor access, what kind of outdoor access, and for how long, are not answered.

Brick decided to try his hand at free-range turkeys. He would continue to confine his egg-laying hens to wire cages, but his turkeys would live in comparative ease. It did not strike him as ironic to treat two similar kinds of birds so differently. Both were business decisions devoid of emotion.

Free-range turkeys, Brick came to conclude a decade later, were among the best gambles he'd ever taken in his life. They were akin to a gushing river—a swollen, steady, dependable income stream. He no longer even bothered to mention egg-laying hens on his business card—just turkeys. Revenues from them were high enough that he renovated and expanded his house. He acquired more trucks and tractors than he could keep track of. He became more financially successful than anyone had ever predicted.

Personally, however, Brick measured his success from the perspective of the "pot." At one time, when his house had been a fraction of its present size, all the Robertses had shared one bathroom. The arrangement had had its challenges, but also, to Brick's mind, its advantages. "When you're on the pot and someone really needs to go," he explained to me, his eyes twinkling, "you have bargaining power." It was that same intuitive understanding of demand and supply that had drawn Brick to free-range turkeys.

From the beginning, Brick's free-range turkey decision—like the Millers' organic dairy decision—had been about forecasts of income, not onsets of ethics. To matters such as the broken turkey fence, Brick paid no heed, for they made no difference to his bottom line.

The labeling system today is not much different than that described in *The Jungle* a hundred years ago. "All of their sausage came out of the same bowl," wrote Upton Sinclair, "but when they came to wrap it they would stamp some of it 'special,' and for this they would charge two cents more a pound."

"Most free-range farms are bullshit," Brick told me. "Most free-range farmers give their chickens and turkeys just a small patch of dirt. It's so small that the buggers don't even bother to use it! Ya won't see a single bugger outside! It's all a scam."

But Brick's official promotional line on free-range was contrary to his personal viewpoint. "We work with Mother Nature, not against her," he would say, straight-faced, when asked about free-range in a professional context.

The three men attired in identical black leather jackets looked like a daredevil motorbike gang, if not for their baseball caps. One cap promoted John Deere, a tractor and agricultural machinery maker, another heralded Bunge, a grain and oilseed corporation, and the third declared, simply, OLD GUYS RULE.

The three men were seated around a table in the same hockey restaurant-bar of my dinner with Nick Roberts. They spoke of Brick, whose fellowship they'd anticipated that evening for their weekly drinks routine, but who'd sent in his stead an unexpected replacement: me.

Earlier in the evening, in keeping with his career and his custom, Brick had slipped on a baseball cap sewn with the image of a chicken. He'd shrugged on a black leather jacket and he'd popped into his faded denim shirt pocket a new cigarette pack. But he'd decided at the door to stay home because he suffered from "a bitch of a headache." He continued, "I'll be drunk as a skunk by the time yer back! Make sure ya pay for all her drinks, Paul!"

Paul and I sat at a small table with Lawrence and Robbie. Fifty-four-year-old Lawrence was towering, taciturn, and commanding, like an evergreen fir. The orange hue of his ponderous handlebar moustache often prompted Brick to ignore his strong Dutch accent and to exclaim "Lawrence is Irish!" Forty-five-year-old Robbie was a father of three girls and was "a good guy," as Paul described to me. I found Lawrence to be too solemn, but I liked Robbie, finding him to be merry and agreeable.

"Did you go to school?" Robbie asked me.

"I did."

"Where'd you go?"

"Dartmouth, in New Hampshire."

"Never heard of it. But I can tell you're educated. What did you do with your degree?"

"I worked on Wall Street."

"I've never met anyone who's worked on Wall Street! How'd you like it?"

"I liked it a lot."

"So, why'd you come out here, to a place like this, to hang out with people like us?"

"I came to volunteer at a dairy farm, then I met Brick Roberts."

"How'd you like Brick?"

"He's great. He's very nice to me. I like him a lot."

"Big change out here from the city, huh?"

"I feel like I'm in a different country."

Robbie laughed.

Over the course of the night, Paul, Lawrence, and Robbie drank a pitcher—each—of Bud Light. Our conversation, which started off well, became rowdy, drunken, and absurd.

Robbie didn't like the idea of living in the U.S. "When you get a lotta Americans in a room, they get really *LOUD*," he hollered. "I don't know how you survived in the States—they're so *LOUD* over there."

Paul responded by calling Robbie a "dough-head" and threatening to yank down his jeans when he stood up. In the meantime, he satisfied himself by smacking off Robbie's and Lawrence's baseball caps. Robbie and Lawrence in turn cuffed off his.

Upon retrieving his cap, Paul dragged his chair closer to mine and dropped a heavy, hairy arm around my shoulders. The clutch tightened and I fled to the ladies room to escape it, recalling what Nick Roberts had said to me the other night: "Uncle Paul drinks a lot. You don't want to see him drunk."

When I returned to the table, Paul was, thankfully, diverted by the prospect of pulling down Robbie's jeans. I asked Lawrence if I could visit his farms.

I could just as well have asked him if I could take his wallet. "I don't let people inside my farms," Lawrence said coldly. "Robbie, Paul, Brick, none of them have been inside my farms."

An uncomfortable silence ensued. More beer was consumed.

"But I don't know about you," Lawrence said minutes later, twirling his moustache. "Brick likes you. . . . I'll think about it. Don't go to Brick's egg or turkey farms tomorrow morning. Come to my house for coffee at ten."

Laid of long, glossy, butter-yellow logs, Lawrence's house was made for magazines. It was a broadcast of the best of the country, a combination of formality and simplicity—the most magnificent house I'd ever seen.

The living room was supported by strong, stalwart log pillars and accented by a dramatic gray stone fireplace that rose brick by brick to the soaring ceiling. The kitchen was designed with brand-new black and white kitchen cabinets whose modern chessboard design created a stylish contrast to the honey-hued rusticity of the rest of the house. In the kitchen lay a black dog bigger than a wolf but balmier than a lamb, his belly serving as pillow to the head of a golden, glowing, giggling girl. The rest of her twelve-year-old form lay stretched out on the cold floor, as the dog lay comfortably on his beige bed placed permanently next to the kitchen table.

It was at the kitchen table that Lawrence's wife, Alva, could be found every morning at ten. First, she buzzed around it like a bee—setting down silverware, a pot of steaming coffee, and plates of browned bread—then she settled suddenly into a chair, where, pale blue eyes flitting to the door, freckled hands patting down a cropped cornfield of yellow hair, she waited for her husband.

At ten past ten, Lawrence Smit marched in with his crimson-haired son. Both men were stained with the unmistakable stench of piles of feces from piles of pigs. The stink screamed that the house's grace grew out of griminess, its beauty out of brutality, its affluence out of affliction. It swallowed all appetite and conversation. The Smits could not separate their family home from their animal warehouse even during their break for coffee.

I suddenly came to loathe the house that I'd come to love a moment ago. Certain forms of enjoyment are dependent on drinking a drug of ignorance. Where suffering exists, it's always there in the shadows, ready to strike at a moment's notice. Our method of acquiring wealth says more about us than the wealth we acquire.

Alva apologized profusely for the smell, her face reddening with embarrassment.

"We have a pig farm and a turkey farm," Lawrence told me, his voice stiff as a stick without the softener of beer. "Alva will show you our turkeys. As for our pigs, our pig farm is just like Charlie's. But I can't let you see it. With pigs, there's a high disease pressure because there are so many."

After a quick breakfast, Alva and I proceeded to her garage, where she requested that I relinquish my boots for biosecurity. The cost of the renunciation was substantial—to her. I happily replaced my cheap black boots with her twelve-year-old daughter's plush, pricey, satin-soft beige boots.

We drove in Alva's black Audi to a set of long, rectangular buildings, and entered one of them. Its entrance was tidy and high-tech, gleaming with red and green knobs and buttons. There, Alva offered me yet another pair of boots—knee-high black boots identical to those at Charlie's pig farms.

"Our big fear is bird flu," she explained in her Dutch accent. "Everybody is really scared of it. We have to be really careful. That's why we tell our electricians don't go to other barns the same day you come to our barns. That's why my husband told you yesterday don't go to Brick's barns this morning before coming here."

I found it ironic that outsiders are considered dangerous harbingers of germs when it is factory farms themselves that are filthy. It's as if the definitions of "cleanliness" and "dirtiness" have somehow gotten switched in the agricultural dictionary. Either way, the primary method of preventing outsider contamination is no more significant than making people change their shoes—equivalent in its effect to protecting oneself from a storm by putting up a curtain.

Alva opened the next door, onto six thousand turkeys.

The turkeys—all of them males, called "toms"—were sixteen weeks old and less than a week away from slaughter. They stood nearly two feet tall, reaching my mid-thigh, forming an infinite, uninterrupted white sea of handfan-like tails and fluffed-up, manure-stained feathers.

Above the feathers bobbed heads of one of three colors: white, indicating excitement; blue, indicating fright; and red, indicating either a desire to attract females or an inclination to fight. The color of a tom's head changes according to his emotional state, a fact that is not just fascinating but also indicative of emotion in turkeys (and, more generally, all birds).

A trilling twitter of sound, called a "gobble," rippled through the ranks of toms regularly, making Alva giggle with pleasure. Ammonia crackled through the air like electricity, assaulting our lungs, abrading our throats, stinging our eyes. Our rapid blinks and tears could not battle it, for it steamed off a caked, crusty carpet of two months of dung.

The Smits' turkey farm was exactly like the Robertses' turkey farm—it was, in fact, precisely modeled on it—but the Robertses' turkeys would be marketed as free-range, based on outdoor space they wouldn't have. Lawrence and Alva Smit were contract growers for Maple Leaf Foods, a multi-billion-dollar Canadian food corporation for which they grew close to ninety thousand turkeys a year.

The toms followed Alva and me as we ambled about aimlessly. "They follow me because they think I am their mother," Alva informed me with pride.

Alva and I knew without a doubt that it was the same large troupe of groupie toms behind us because of their injuries. One tom was recognizable by the stub that would have been his snood—it held the flinty black color of old blood. Another was identifiable by a large, round patch of bruised, featherless skin. These two toms were healthier than many, though. They were not in the cardboard-bordered "sick pen," which contained a hundred or so toms in varying states of illness and affliction.

"Lawrence comes in here one or two times a day for fifteen minutes to check things and take out dead toms," Alva said.

But Lawrence had missed many dead toms because of the brevity of his visits, and Alva decided to be helpful to him by collecting them herself. They were easy to find, being visible every time the hordes parted.

The first dead tom Alva retrieved had deteriorated to carrion—red, pecked, plucked. "It probably had a heart attack," she speculated. "Most of the time, heart attacks is why they die. Breeders breed them to grow so fast that their heart and vascular can't keep up. The body weight of turkeys increases three hundred times between birth and slaughter. In their last few weeks, they put on two hundred grams of weight every *day*. When we started with turkeys ten years before, we had to keep them for seventeen weeks, and they became fifteen kilograms [thirty-three pounds]. Now we keep them for sixteen weeks, and they become one kilo heavier—sixteen kilos [thirty-five pounds]. Their genetics are crazy."

Today's turkeys are called Broad Breasted White, and, as suggested by their name, they are bred for an unnaturally large breast. This is because the breast is the most profitable part of the body—the part in highest demand by consumers, who are unaware of the welfare repercussions of their taste preferences. The large size of modern turkeys also means that they are no longer able to mate naturally, as the tom is too heavy to mount the female. Consequently, they are artificially inseminated: male breeding toms are "milked" for semen collection, and females are then inseminated by syringe.

Every year, the President of the United States ceremonially "pardons" a turkey, giving the animal a reprieve from slaughter. Thanksgiving is "one of the worst days of the year to be a turkey," President Obama jovially noted in his 2011 turkey pardon. He continued that Liberty, the turkey he pardoned, is probably "the luckiest bird on the face of the earth."

Not quite. Most pardoned turkeys tend to drop dead in under a year because of their misshapen body structure.

Watching Alva collect dead toms, I realized that farmers today spend more time removing dead animals than caring for living ones. Different kinds of farm animals die of different causes: turkeys die of heart attacks, piglets of castration and tail-docking, egg-laying hens of hangings, blowouts, and caging injuries. These harsh forms of deaths are considered a standard cost of doing business.

Alva collected a second casualty of "crazy" genetics, flattened into the floor of feces by flocks of feet. A short distance from him, she found a third one as well—feathered, fresh, a recent expiry. Alva added the two carcasses to the one in her hand. Her face flushed from the weight of the toms and the effort required to respire in the ammonia air, she strode to the door, the three dead heads bouncing behind her like deflated squash balls.

She dumped the toms at the door, next to another pile, this also of three toms. Two of the toms in the pile were dead; the third was dying but not dead, living but not alive, his skin sparkling scarlet with blood. Alva kicked him.

He turned his head. She kicked him again. He squeaked. She kicked him harder. He shrieked. Her boot rose again—"Why are you kicking it?" I cried.

"I want to see how sick it is," she answered matter-of-factly. "Whether it will live or die."

"It will die. Lawrence thinks so, too, otherwise he wouldn't have put it here, beside the door, with the dead ones."

Alva stared at me like she'd never heard such logic before, her eyebrows ascending into her blond bangs. With a shrug, she kicked the tom again.

Quavering to his feet, he budged one small step away. Then another small step. He collapsed with a thump onto the dung.

"You are right," Alva concluded. "It will die."

We left him to die.

CHAPTER FIVE

CRANK CHICKEN

THE END OF THE BEGINNING

For how mammoth a tractor is on the outside, I assumed it would be spacious on the inside. It was not.

As Nick Roberts gave me a tour of his family's property—400 acres of bush, 800 acres of fields, the family house, the turkey buildings, the silos and feed bins, the egg-laying hen factory, Paul's abode, the ramshackle shanty housing the hunting dogs, the wood cabin for hosting summer barbeques, the small pond carved in front of the cabin for fishing—Nick's wide, warm thigh pressed against mine on the narrow seat of the tractor. When we turned our faces toward each other, our lips were separated only by inches, our breaths mingling in the air.

The forced intimacy of the tractor seat was why, when we returned to the Robertses' home, I opted to sit across the coffee table from Nick. He stretched out on the couch like it was a bed, then he lifted the family dog Lizzie, who'd just walked in, on to his chest. Cradling her in his arms like a lover, he instructed me, "If Lizzie goes to the door, or if she even looks at it, make sure you open it so she can go out."

With that, Nick covered himself and Lizzie with the red blanket at his feet and fell asleep, his arms wrapped tenderly around her long black back. When he started snoring, lightly, rhythmically, as a train's engine hums, Lizzie carefully disentangled her limbs from his, crept off the couch, and trotted to the door. I rushed to open it, per Nick's mandate.

Lizzie was such a cherished member of the Roberts family that they'd even prepared a birth certificate for her. In a silver frame in the kitchen among family photographs, Lizzie's certificate stated that her full name was Elizabeth Roberts and that she was born on the sixth day of September with seven brothers and sisters, to a dog called Trixy.

Lizzie gave the Roberts family nothing—not flesh, nor milk, nor eggs, nor even assistance in appropriating flesh or milk or eggs by herding or guarding or shepherding—but she was given everything. She served no function on a farm, but she, like all the other farm dogs I'd seen so far, was permitted to roam free, indoors or outdoors, as she pleased. Farm animals, in contrast, were rewarded for their flesh, milk, and eggs by confinement to crates, cages, and chains. One sort of animal was exploited to the utmost and the other pampered with no limit.

Lizzie returned, and I ran to open the door for her. She sauntered in without a glance at me and proceeded to her favorite place in the house: the wood stove, which happened to be located near her birth certificate. She reheated her aging bones as she always did, by laying her head directly underneath the furnace, her face sagging with sleepy surliness.

When Nick's snores reached a crescendo, Brick Roberts burst into the living room. His torso was naked. His belly was a red rock matted with a moss of gray bristles. From among the bristles peeked out his belly button, looking like a snarled, spiral eye. He placed an arm around my shoulders, kissed the top of my head, then plopped down next to me on the sofa and began scratching his back with a hand-shaped back-scratcher. "My back is sore from the snowmobile," he explained. "Did he jump into bed with ya yet?"

"What?!"

Brick was looking at Dexter. Not Nick. "Oh, Dexter, right," I recovered. "No, he hasn't. I've been keeping my door closed. . . . Sorry."

Brick believed that his dark, drowsy cat Dexter led a secret life, that though he appeared during the day to be surgically attached to the armchair in the living room—not coincidently the most comfortable chair in the house—he prowled the house at night. To assist in Dexter's nocturnal adventures, Brick insisted that all bedroom doors be left open all night. I always closed my door, though (feeling myself to be so much a part of the Roberts family that I made my own decisions against Brick's express instructions).

"Once, I took Dexter off his chair and I sat on the chair," Brick reminisced. "But Dexter didn't like that. No, he didn't. He scratched me up! I was bleeding!" His voice rising with pride, Brick indicated the gashes, now blanched, that Dexter had chiseled into his boulder belly. He then pried the twenty-pounder off his armchair and deposited him onto my lap.

As Dexter dug his claws into my pants, I realized that it was strange the Robertses had not declawed their cat but had declawed their turkeys (in addition to de-beaking and de-snooding them). Perhaps it was because declawing a cat is considered cruel, and declawing turkeys acceptable. Cats fall under societal norms and turkeys under agricultural, two realms that, as I was finding, have nothing in common.

When Nick eventually awoke, Brick stowed his back-scratcher, donned his faded denim shirt, and switched on the television. Both father and son shunned *The Office*—"I'd rather poke my eyes out than be in an office," Nick explained—and settled on *1,001 Ways to Die*. Despairing of watching deaths from causes like cock-fighting, leaf-raking, and electrocution, Jane arranged on the kitchen table a board game that she'd played when she was little, and that she'd taught her children to play when they were little.

Brick and Nick grudgingly shut off the television and joined Jane and me at the table. Brick and Jane formed one team, and Nick and I another. He and I high-fived each other as enthusiastically as soccer players every time we scored, even though Jane was clearly the best among the four of us.

That evening—of miscellaneous television, a childhood game, and, later, dinner cooked by Jane—was my favorite with the Roberts family.

"Since you've decided to return home tomorrow," Nick told me, "let's leave together early in the morning, so we can spend the whole day together in the city."

"Okay," I agreed. But I still intended to leave without him.

The buildings stood equidistant from one another like sentries, attired in a clean paint of white hemmed with green. They were as long and warehouse-like as all the other animal facilities I'd seen, but they were taller—two stories tall this time.

Inside the sparse, brightly lit, laboratory-like entrance of one of them stood Terry, a twenty-two-year-old with green eyes, brown hair, and a round face made angular by means of a precisely pruned goatee. Terry knew Nick from school, and among his fondest childhood memories was eating "Uncle Brick's" turkey sausages as a teenager. That was why Terry had acquiesced immediately, if reluctantly, to Nick's request for a tour, made on my behalf. Our meeting time was ten o'clock on Friday night, because Terry had claimed to be unavailable earlier.

Along with more than a hundred other contract growers, Terry and his father together grew more than half a million broiler (meat) chickens every year for Grand River Foods, a processor that supplied Walmart, Safeway, Pizza Hut, Subway, and other large stores and restaurant chains.

Dropping four transparent covers on the floor, Terry directed, "Use these for biosecurity." I began wiping my boots on the covers thoroughly, as on a rug, until Terry's scathing glance stopped me in my tracks. A quick look at Nick showed me that I was supposed to wear the covers *over* my boots, not wipe my boots on them. Terry's first impression of me—that I was a moron—only worsened over the course of the night.

"You guys remembered what I told you, right?" Terry asked us nervously. "You're not wearing the same clothes and boots here that you wore to other farms, right? Right?"

"Right," Nick replied.

We'd taken great care to follow Terry's clothing instructions. New to animal agriculture, I'd brought only a few changes of clothes with me in my suitcase for my stay with the Robertses, not realizing that even momentary ammonia exposure would leave them tainted with an unendurable stink. It was for that reason that I was presently wearing Jane's clothes and boots, which she'd been gracious enough to lend me. They were extra-large and I was extra-small, but they were holding up, at least for the moment.

Terry exhaled with relief at Nick's reply, then inhaled deeply, like he was trying to stockpile oxygen in his lungs before a deep-sea dive. As he opened the next door, it was not with fanfare but with an apology. "Sorry it stinks in here."

The stink was of 40,000 chickens, half of them in front of us, and the other half of them above us, on the second floor. The chickens before us were colored like tri-flavor vanilla, chocolate, and strawberry ice cream, their white feathers smudged brown, their featherless areas pink. All of them could be smelled, but only the closest of them could be seen, for the place was as dark and eerie as a cemetery. It was also foggy, with wisps of cool winter air floating in—white, ghost-like—through narrow vents along the top of one wall.

"Could you turn the lights on?" I asked Terry, covered with goose bumps. "Just a little? So we can see better?"

"No. The lights have to be very controlled, so the chickens gain weight but don't get heart attacks. When the lights are on, the chickens are awake and eating and gaining weight. When it's dark, they stop eating. We don't want them to eat too much, because their genes are weird. They grow too fast, and their heart and legs get fucked up. They get heart attacks. So, we need to control how much they eat, and we control it with the lighting levels."

Chickens and turkeys today are, in a sense, like balloons, except that they expand not with air but light. If they enlarge too fast, they explode—or, rather, implode, collapsing on painful, broken legs. Extreme genetic selection, accelerated by artificial insemination, has created farm animal breeds today that yield far more meat, milk, and eggs—while eating far less—than they ever have. The most astounding genetic changes have been those of chickens. In 1925, chickens reached a weight of two and a half pounds in sixteen weeks; today, they reach a weight of almost six pounds in six weeks (while consuming less than half the feed per pound of weight gained). It's miraculous but torturous.

"Today's poultry chicken has been bred to grow so rapidly," states animal expert Dr. Temple Grandin in *Animals in Translation*, "that its legs can collapse under the weight of its ballooning body. It's awful. Darkness slows down the baby chick's growth just enough to prevent this from happening, so getting those lights off [at night] is important, because lameness is a severe problem in chicken welfare. I've been to farms where half of the chickens are lame." As genetic selection continues, this problem, which Dr. Grandin terms "biological system overload," will only increase.

Whether or not they were lame, Terry's chickens were clearly indisposed to movement, sitting as committedly as if they'd been glued to their dung on the floor. There was an irony in this: Brick Roberts's egg-laying hens *wanted* to move, but couldn't, on account of their cages; Terry's broiler chickens *could* move, but didn't want to, on account of their genetics. (While egg-laying hens in North America are generally confined to cages, broiler chickens are not, because of the industry's fear that cage wire will cause breast blisters and lower the price of the flesh in the market.)

The breed of Brick's egg-laying hens had a poetic, pastoral name, "Rhode Island Red," arising from the name of the state where the birds were first bred. Terry's broiler-chicken breed, in contrast, among the most popular broiler breeds in the world, sounded more like a computer program than an animal: Ross 308. The name was fitting, however, because it was as devoid of grace and color as the ballooning breed itself—and also because Ross 308 chickens *were* a computer product. In the words of Aviagen, the corporation that designed them according to strict calculations of "growth rate" and "feed efficiency," Ross 308 broilers meet "a broad range of end product requirements" and provide "integrated operations with the perfect balance of breeder, broiler and processing performance."

Ammonia rising from manure pricked our eyes and scratched our throats, and we coughed and blinked like we were trapped in a dust storm. Terry's face whitened to an ashen gray, and he hacked like a lung cancer patient. He squinted through the fog at the door, like a criminal plotting his escape.

Nick Roberts dug the tip of his plastic-covered boot into the excrement, wondering how far his foot would have to penetrate to reach the floor to the other side. "That's not just shit," Terry notified him hurriedly. "There's lots of beetles in there. We have a big fucking problem with beetles."

But in the dark mist, the beetles blended into the dough of dung like chocolate chips into a black forest cake—they were impossible to locate. All that we could distinguish in the field of feces was spilled feed, forming yellow halos around the feeders. "We feed the chickens corn, soy meal, and meat meal," Terry said. "The meat meal comes from pigs."

I hadn't known the meaning of "meat meal" until Brick Roberts had explained it to me. In his egg-laying hen factory, a feed pellet on his callused palm for demonstration, Brick had said: "My pellets aren't like most pellets. Most pellets have slaughter by-products in them. The slaughter people take the slaughter leftovers and 'render' 'em. That means they grind the by-products up into meat meal, or bone meal, or some other kinda meal.

"Then, what they do is, they *add* that rendered stuff to the feed pellets with the corn and the soy. Ya can't buy separate corn pellets and soy pellets for chickens; all the pellets have everything in them together, including the by-products. Feed producers sell chicken feed like this so chickens can't choose what they eat. They *have* to eat slaughter by-products, 'cause they're in the pellet."

Terry's chicken feed, in contrast to Brick's, was in the form of powder, not pellets. Terry grew his own corn, purchased soy meal and pig meat meal elsewhere, and mixed them in with the corn. The three kinds of granules then surged through the pipes and dangling red feeders simultaneously but separately, as distinct particles. This meant that, unlike most chickens, Terry's chickens *could* pick through their feed. They could express their eating inclinations through their eating order.

"They pick out the corn and soy first," Terry said.

Of course they did. It's unnatural for chickens to eat pigs. Up to a quarter of the diet of Terry's chickens, however, consisted of pigs. (The peak of profitability would be to concoct, within closed walls, a closed food chain: to get animals to eat what they excrete. Such a feces-focused food chain would be an agricultural breakthrough. Meat would be cheaper than ever, hardly costing more than manure.)

"It's not right to feed animals slaughter by-products," Brick had continued to tell me in his hen factory. "Chickens aren't supposed to eat other farm animals. That's why I make sure my pellets don't have by-products. My feed costs more than other farmers' feed because my hens and turkeys are 'vegetarian-fed.'"

Only a paltry percentage of chickens and turkeys in the United States and Canada are vegetarian-fed, however. The great majority are regularly fed a diet involving animal by-products. "The poultry industry in the United States has a long history of using rendered products in its rations," explains a paper titled *Rendered Products in Poultry Nutrition*. "Rendered fats are generally lower in cost than vegetable oils such as soybean oil, which is used substantially in other countries. . . . Combined, these [rendered] products can be used to provide a substantial cost savings to the poultry industry and use of the products is quite high by the industry. . . . Strong utilization of these products by the poultry industry is the norm and is expected to continue into the future."

"For chickens, you look at profit per kilogram," Terry wheezed to me. "We do a least-cost analysis. That means we feed what costs least. By-products cost least, so we feed by-products. We're here to make money. It's a business."

In Terry's business, meat meal came from Rothsay. Among the Robertses and their friends, the name Rothsay was mentioned so regularly that I initially assumed it was an important neighboring town. It wasn't. Rothsay was a renderer, the largest in Canada. It belonged to a food corporation, among the largest in Canada, multi-billion-dollar Maple Leaf Foods. (In late 2013, though, Rothsay was purchased by a corporation called Darling Ingredients, a publicly traded, Texas-based renderer operating more than 200 facilities worldwide.)

The name Maple Leaf was mentioned in the Robertses' community even more regularly than Rothsay, since everybody—*everybody*—worked with Maple Leaf in one capacity or another. Terry purchased pig meat meal from Rothsay. Lawrence Smit grew turkeys on contract for Maple Leaf. Brick Roberts had his free-range turkeys slaughtered by Maple Leaf. It was Brick's meeting with Maple Leaf earlier during my stay that had left him with renewed rancor for the "high-volume, low-margin" model of agribusiness.

The way it worked was that Maple Leaf supplied animals to contract growers to, as their job title implied, grow on contract. Then, when the animals had attained the requisite weight in the requisite time according to the requisite instructions, Maple Leaf slaughtered them. Rothsay and its 500 employees handled the indigestible, unpalatable, contaminated, and condemned animal parts from the slaughter: the by-products.

They cooked, crumbled, and crushed hundreds of millions of pounds every year of heads and tails, livers and lungs, bones and kidneys, feathers and blood, and skins and intestines. The by-products made from these body parts included: meat and bone meal, to feed to chickens, turkeys, and pigs; pork meal and pig offal meal, to feed to any farm animal at all; and blood meal and feather meal, to feed to the most herbivorous of farm animals, dairy cows. Dairy cows in the United States can end up eating one pound of other animals per day.

Rothsay did not limit itself to rendering animals killed at slaughter plants. It rendered also animals that died *before* slaughter, *at* farms—animals that were ill and diseased—and it sold their ground-up body parts to contract growers to feed back to their living farm animals. Rothsay even offered "on-farm collection" services to pick up animals that died at farms.

For ease of the collector's weekly visit, Lawrence and Alva Smit kept five big blue bins full of dead turkeys just beside their log lodge, even though the bins degraded the grandeur of the lodge. Brick Roberts, in contrast to the Smits, had no use for any kind of "on-farm collection." Because he shunned slaughter by-products, he shunned also any contribution to them. Paul dropped egg-laying hen and turkey carcasses into a deep freezer, then buried them when the freezer became full.

Terry looked sick, his face drained, his eyes glassy. "I can't fucking take this," he croaked. "I need to get out *now*!" He ran out.

In the entrance where he'd waited for us earlier, Terry keeled over. His hands on his knees, he heaved for breath like he'd just escaped a fire.

"I have asthma," he gasped. "If we'd been in there ten more minutes, the ammonia would've killed me. I always wear a mask when I go in there. I didn't this time 'cause I didn't know you'd want to fucking *stay* in there."

He glared at me balefully. "I'm sorry," I said. Shaking his head, he glanced away like he was too hurt to respond.

When he recovered, Terry opened the door of a small cabinet on the wall. It was filled with masks of all kinds: cloth masks, surgeon-style masks, firefighter masks, masks that swallow your whole head like a bear swallows a fish. Terry could start a Halloween rental store on the side.

"If I didn't have asthma, I wouldn't wear a mask," he coughed. "Then I'd have bigger problems than asthma because of all the ammonia. Asthma sucks but it saved my life 'cause it forces me to wear a mask."

The ammonia in factory farms was *that* intolerable, that it actually made Terry glad of his asthma.

"The air is awful for you," Nick Roberts agreed. "I've been telling Uncle Paul to wear a mask for years, but he doesn't, and he has this crazy cough now."

Terry's masks, I noticed, were pristinely clean, like china in a glass cabinet. "I rarely wear them," he explained. "I come in here only when there's a problem."

"But how do you know there's a problem if you don't come in?" I asked.

"My phone," Terry said, brandishing it in front of my face.

"I don't understand."

"How long has she been with you guys?" Terry asked Nick incredulously, referring to me like I was absent. "It's hard for you to understand this stuff," he said, turning back to me, "but, for example, temperature. I've set the temperature in the barns to be between 65 and 85 Fahrenheit. If it goes below 65 or above 85, I get an automatic call on my phone. Then I come in and take a look. Otherwise, I don't need to come in."

"But don't you need to check on things even when there isn't a problem?"

"I check on things with my phone," he said impatiently. "I'll show you, so you actually get it. . . . See, I'm dialing Barn 3 right now. That's not this barn, it's another one."

Ring, ring, ring, ring. . . . "I've set it to ring a lot, so people give up if they call by accident," he explained. Barn 3 eventually received the call, and a female machine voice began providing status updates, first for "Zone 1," then "Zone 2," then "Zone 3". . . .

"There are eight zones," Terry elaborated. "Zones 1 and 2 are physical areas of the barn, and Zones 3 to 8 are the different equipment. I get a status on all the zones by calling in."

With the zones, and the phone, and the robotic tone, the chicken operation was like a futuristic science fiction film—a version of *The Matrix*. Factory farms are so automated that people don't even need to be present for supervision. It is difficult for people to even *survive* in such toxic environments, in fact, Terry being a case in point.

The phone trend was not limited to Terry, though. The number of people working on farms has fallen everywhere in the last half-century. In the U.S., just one worker is often responsible for thousands of pigs and hundreds of thousands of chickens and egg-laying hens. "The United States now has more prison inmates than full-time farmers," states journalist Eric Schlosser in his book, *Fast Food Nation*.

Fifty years ago, animals were fed and tended by people, and so conditions at farms had to be bearable enough for people to work there. Today, these tasks are performed by machine and phone. Smartphones turn lights on and off, lock and unlock buildings, and track animal feed and water intake. Farmers can choose between apps with names like Cattle Breeding Calculator, Pro Dairy Event, and Pork Production. "Mobile monitoring of all aspects of farming can only grow," predicts a report on phone farming prepared by a mobile learning company.

There's nothing wrong per se with technological replacement of people in any industry. The trouble with the phone trend in agriculture is more subtle. If contract growers no longer have any reason to enter their operations, they no longer have any personal incentive to make conditions more livable for their animals.

Terry invited us to his house for beer.

The time was close to midnight, and I demurred. Nick, however, agreed.

"Just in case it comes up, I need to tell you something," Nick mumbled to me in his truck on the way to Terry's house. He paused, then rambled onward like a freight train, in the tone of confessing a sin. "I used to go out with Terry's girlfriend, Kelly! But that was when I was in the eighth grade, and she was in the ninth grade. I'm totally not interested in her now, you know! But that doesn't mean I don't like older girls!" he clarified, remembering that I was two years older than him. "I like older girls—a *lot*. . . ."

With its new houses and sleek cars, Terry's neighborhood was more chic-suburban than rural. So was his home, with its tiled floors, granite kitchen counters, and a large, flat-screen television, which was blaring *Crank 2: High Voltage*. Terry's girlfriend, Kelly, and two friends—a solemn, rail-thin dairy worker and his pretty, bubbly girlfriend—were watching the action/sex flick intensely.

"Aww, that's *so* sad," moaned the dairy worker's girlfriend.

A man in *Crank 2* was teaching his German Shepherd who was "in charge" by using an electric collar. He pressed a button on his remote control to shock the dog, who whimpered, then barked. Chev Chelios, the lead actor of *Crank 2*, played by Jason Statham, happened to run onto the scene. "How do you sleep at night?" he berated the electrocuter. Liberating the dog and patting his head, he claimed, "I'm from PETA. This is animal cruelty." The electrocuter called him a "liberal freak."

"Will's used electric prods on me a couple of times when we've gone to pig farms," Nick said as he watched the dog-shocking scene. "It hurts like *hell*!"

The rail-thin dairy worker, though, despite his girlfriend's groans, looked as unmoved during the dog scene as if he were watching an advertisement for dish detergent. His dairy farm, like

Michael Miller's, probably hung electric trainers above chained cows.

The dairy worker and his girlfriend left soon after. Nick and Terry stepped out of the house to smoke and to admire Kelly's new car, which Kelly described as "a beauty" and "the love of my life." I was left alone with Kelly, a twenty-five-year-old with short, unwashed blond hair and a syrupy voice like artificial sweetener. She was conflicted about Terry's chicken-growing work on multiple levels: as a girlfriend, as an animal lover, and as a vegetarian.

"Chickens *stink,*" she told me. "I tell Terry not to go into the barns, but sometimes he has to. When he comes back, he smells just like chicken poo. I make him put his farm clothes in the basement, but I can still smell them from all the way up here! I'm allergic to dust. I go into the barns only on the first day the chicks come. Then, me and my friend run around and grab as many as we can. We put necklaces on two of them and we call them our pets!"

Adjusting her black Dolce & Gabbana eyeglasses, Kelly arranged her arms into the position of cradling a baby. "I love chicks. They're *so* cute! I just *love* animals. I'm a vegetarian."

I realized that I'd hardly ever heard about vegetarianism more than I had in this animal farming community that I'd happened to stumble into. It can't be a coincidence that some of the people closest to the system of meat production are boycotting meat.

Kelly showed me a photo on her phone of a daffodil chick resting on the palm of her hand. "I want to keep a pet chicken in the house, but Terry won't let me."

Nick and Terry returned then. "Terry, can I keep a pet chicken in the house?" she implored with a pout.

"No fucking way!" came the reply. Kelly unfurled her middle finger at him.

Nick and Terry joined Kelly and me around the kitchen table and shared stories about drinking and fighting. "This one night,

I'm taking a leak behind a house," Nick recounted. "Out of nowhere, this old fucker comes up behind me and socks me in the jaw! . . ." Terry started relating his own tale of getting punched—"I was at a bar, and this fuck-face came up to me . . ."—but he left it incomplete, for *Crank 2* ended and pornography started.

Terry rushed to shut off the television. Nick interpreted the move as a cue to leave.

I was dreading this moment, when Nick and I would be alone in his truck, and I would lie.

Hesitant to lead Nick on by spending more time with him, I wanted to take the bus home the next morning. Nick, however, was equally determined to drive me in his truck, proving immune to arguments such as "I love taking buses!" Earlier in the day, I'd given up on the prospect of leaving alone, but in the evening I'd had an idea, inspired by Nick's impassioned opinion while we were watching television: "I'd rather poke my eyes out than be in an office."

"I'll spend tomorrow at my sister's office," I told Nick now in his truck. "It'll be *really* boring. Trust me, you don't want to come. I'll just take the bus back in the morning."

Nick's bottle of beer froze midway to his mouth. "But isn't tomorrow Saturday?"

I should've thought of that. "Right," I stammered. "But I need to go to her office anyway."

"Why?"

"Maybe I'll help her with her work . . . ?"

It was that final question mark, dangling in the air like a noose, that told Nick the office was a lie, an excuse. His face turned to stone. "You're right," he said tightly. "I don't want to spend my Saturday in an office."

Nick smoked and drank in silence. He did not invite me over to his place for a drink. He did not smile or wave good-bye when he dropped me off at his parents' house. I never saw him again.

My stay with the dairy-farming Millers had been an exercise in hostility, and it had led to my self-conscious feeling that I was a bad guest, perhaps a bad person. It had also led to my developing two fears during my stay with the Robertses: that I'd run into the Millers, and, even if not, the Robertses would dislike me as much as the Millers had.

Neither scenario had occurred. "Yer a part of the family," Brick had told me repeatedly. For reasons unknown to me—and perhaps even to him—he'd trusted me with the keys to all compartments of his life. He'd shared his home, his family, and his friends with me. His friends had in turn, because of their regard for him, shared their homes, their families, and their friends with me.

In his closed agricultural community, Brick's endorsement was like a VIP access card. Without the card, all doors were closed and all protocols followed. With the card, many doors were opened and protocols ignored. During my residence with the Roberts family, I'd stepped into three houses: Charlie's wood cabin, Lawrence's log lodge, and Terry's suburban home. In contrast, in the whole of the two years I'd worked on Wall Street, I'd entered just *one* colleague's apartment, just once, for a house party.

In this faraway farming landscape, I could not help but conclude that people were genuine, that they meant what they said and they said what they meant. And that was why I could not refuse Brick when, after I stayed with him, he wanted to stay with me.

"It's Uncle Brick," Brick hailed over the phone two weeks after I left. "I'm giving ya a ding to tell ya I love ya! I hope yer happy, healthy, wealthy, and wise. I'm in my office with yer boyfriend right now, and we're drunk as a skunk! We're going to the farm show next week! The show's kinda near where ya are, and ya should meet us there."

"Sure. I've never been to a farm show before. What should I wear?"

"I wear a tuxedo whenever I come in to the city for meetings. I'm not gonna wear a tuxedo for the show, but I am gonna dress nice! Oh, and I'm coming with yer boyfriend and twenty-three farmers. We're going to be drunk as a skunk, and we're staying at your apartment that night! I'm just kidding. There won't be twenty-three of us staying with ya; there'll be thirty-five of us!"

Thirty-five farmers? In *my* apartment? Drunk as a skunk? No. *No. NO.*

But Brick sounded so full of cheer, so buzzed with beer, that, squirming like a slug, I muttered "Okay. . . ." I really hoped he was kidding, but with Brick, you could never tell.

"I'll give ya a ding when we get to the show! I love ya."

Brick paused, waiting for a reply. "I love you too," I mumbled.

Taking Brick's cue, I showed up at the farm show wearing a button-down black shirt and a knee-length black skirt. Such an ensemble was casual for Wall Street—I'd worn a suit to work every day—but the farm show, as it turned out, was a sprawling center of trucks and tractors and trailers, and draining and lighting and painting equipment, and concrete and feed and seed suppliers. Almost all of the thousands of attendees at the farm show were men, and almost of the men wore baseball caps and jeans, and most of them stared at me as strangely as if I were a sow in stilettos. I could just as well have worn a ball gown and topped it off with a tiara.

Brick, however, had not been kidding about dressing nice. He wore a crisp plaid shirt, a brown belt with a bright buckle, and, most noticeably, a pair of glittering, red, one-dollar sunglasses lined with fake diamonds, which hung on his wide neck without need for a strap. Brick had been kidding about thirty-five farmers staying in my apartment. Flanking him were only Lawrence Smit and Robbie, both of whom I knew. All three men had reserved hotel rooms for the night; nobody would be staying with me. I felt like singing with relief.

Brick gave me a rib-crushing hug, then draped his cigarette-scented black leather jacket around my shoulders and wrapped an arm around my waist. "Uncle Paul wanted to see ya, too, but he couldn't come to the show 'cause he's playing golf in Mississippi for four days," he told me. "And yer boyfriend's right over there, beside that solar panel."

Really? Nick was here? At the farm show? This meant that he was probably no longer upset about our last conversation! Pleased, I turned toward the solar panel. . . .

It was not Nick. It was his younger brother, Will.

I felt a surge of disappointment, for I valued Nick's friendship and I regretted how it had ended.

Will, Brick, Lawrence, Robbie, and I strolled about the agricultural equipment together, forming a team as odd as Winnie the Pooh's. The four men were most interested in solar panels. To my surprise, there were dozens of solar panels at the farm show. Some of them were the humble dimensions of home television sets, and others were the towering size of cinema screens.

"I'm putting in more solar panels," Brick told me. "I've already installed some on one side of the turkey barn." I'd seen those.

"I have four panels," Robbie shared.

"I have a couple," Lawrence added.

"How do solar panels work?" I asked.

Robbie detailed the prices, payment structure, and profitability of panels over a full two decades, stating in happy conclusion, "The solar panels will make enough to put my daughters through university."

"But *how* do solar panels work?" I repeated. "*How* do they convert sunlight to electricity?"

Robbie shrugged. Lawrence scratched his OLD GUYS RULE cap. "I'm gonna go take a whiz," Brick announced.

Farm animals, I realized in that moment, are, to contract growers, like solar panels. Solar panels are valued for converting solar energy to electric energy; farm animals are valued for converting feed to flesh. All that their owners know about them is their prices, their payment structure, and their profitability—little else. The only difference between farm animals and solar panels is, ironically, the greater freedom of the panels. Panels can see the sun, and, as they trace its passage through the sky, tilt and turn.

There is a further irony in solar panels. Farms of the past were grass-based and, as such, solar-powered by definition: the sun nourished the grass, and the grass in turn nourished the animals living upon it. Today, the sole role that sunshine plays on factory farms is to grow the corn and soy that animals eat. Otherwise, solar panels define the only relationship that factory farms have with the sun today.

The contract growers of today have nothing in common with the farmers of old. Until a half-century ago, farmers raised and reared animals; in contrast, contract growers grow animals to a decreed weight, as one grows grass to a decreed height. The farmer's goal was quality; the contract grower's is quantity. The farmer's focus was animal health; the contract grower's is animal heft. The farmer's farm had been open to the world; the contract grower's is its own closed world of disease, its borders barricaded by biosecurity. Animal use has become synonymous with animal abuse.

On my way out of the farm show, I quietly purchased a pair of second-hand farm coveralls.

On Wall Street, I'd often worked until the wee hours of the morning, but no night at my desk had been as challenging as my stays with the Robertses and the Millers. Animal farms were an assault on all my senses. The sows especially—their head-banging against their crates, their ringing screams, the fecal fog—I could not get them off my mind. The sights, sounds, and smells of factory farms haunted me, flashing like lightning across the storm of my memory. I could no longer sleep at night.

When I'd begun my farm vacation with the Millers, I'd imagined something like a swim about a pool, lonely but leisurely. But a complication had arisen. The pool into which I'd dived had drained into the entire ocean of animal agriculture. The floor had vanished from beneath my feet, and the game had transformed from one of enjoyment to endurance. Should I retreat to shore, I now asked myself, or flounder farther to sea?

The shore was fringed with friendly, familiar fronds—employment, family, money, security, society, city: everything I'd ever known and loved. The ocean of agriculture, in contrast, was cold and callous, unpredictable and alien. The choice between the two options seemed obvious. There were several reasons I should leave the rural route immediately.

First was profession. In my school years, I'd stayed up late into the night learning rather than living, studying rather than partying, an unbalanced type A machine, a battery that needed no charging. Born into a middle-class household, I'd worked diligently and obsessively with the aim of becoming rich one day, fantasizing about all the beautiful clothes and accessories I'd acquire. And then, on Wall Street, I'd finally attained the safe, set life I'd sought. I was able to buy myself whatever I wanted—and I bought plenty. My lunches and dinners were paid for, and so were my cab rides home at night—and so was my life.

But if I wanted to return to Wall Street, I'd have to do it immediately, because plenty of time had passed since my last job, and if any more did, no firm would want to hire me. Pushing open the next agricultural door would be equivalent to slamming shut the Wall Street door with my other hand. But I could hardly bear the thought of it. I feared I would regret the decision for the rest of my life.

Second, probability. Even if I did decide to prick my career like a balloon, there was a high probability that farm doors would remain locked regardless of how long I knocked. Brick's endorsement had been my key, my immunity, but he'd already introduced me to everybody he knew. Going forward, I would be on my own, a novice in an unfamiliar forest.

Third, danger. Classrooms and cubicles are, by their nature, hazard-free. The gravest dangers that I'd encountered at school and work were overdue assignments. Living in the homes of strangers and traipsing about unknown areas, in contrast, are dangerous, foolhardy ventures. One can easily slip through a crack, without anyone ever finding out.

Fourth, driving. The Robertses had been kind enough to truck me around, but going forward, how would I go *forward*? I could not drive, and there were no subways running under corn fields.

Finally, family. My parents had been proud of me for the job I'd had. They would be proud of me no longer. My mother was frantically worried about my unemployed state, though for more reasons than the obvious. "Who will marry you if you're unemployed?" she said. "Don't you want to get married? If you ask me, I think you're ruining your life."

It was for these five reasons that my intellect urged me to leave the ocean of agriculture instantly, even as my intuition pleaded with me to remain, to not conclude my journey until I'd at least attempted to address the crushing problem of farm animal cruelty.

Uncharacteristically, I chose intuition over intellect.

Yes, I would be drowning my career. Yes, I might not be able to enter farms. Yes, I would be risking my life. Yes, I could not drive. Yes, I would be incurring my parents' disapproval. But I saw no real choice left me.

Without solutions, any contemplation of afflictions is like visiting a doctor and departing without a diagnosis. Solutions transform anger to power, chaos to order. I would dive deep into the ocean of animal agriculture, I decided, and emerge from the foaming froth only when my hands held the luminescent pearls of solutions I sought.

CHAPTER SIX

ANIMAL HEAVEN

QUALITY WITH A CONSCIENCE

The motivation behind our farming system is allowing our farm animals to have the best life possible, letting them live in family groups for as long as possible, giving them a stress-free life. . . . Our animals are the reason we farm, so our first priority is to understand them; what does each and every one of them need to reach their full potential?"

This was found on the website of Harley Farms. After the factory farms I'd seen, I hesitated to believe the Harley Farms website. The thoughtful words, the bucolic pictures—weren't they likely to be lies? I phoned Harley Farms to find out.

My conversation with the farmer, Roger Harley, lasted close to an hour. At the end, he extended me an invitation to come stay at his farm and see his ideals in practice for myself. I agreed enthusiastically.

Roger lived in a little village called Keene, two hours east of Toronto, three and a half hours from upstate New York.

The Harley home consisted of two houses affixed together like stones, one white, the other red brick. The houses were originally constructed by settlers close to two centuries ago, but Roger and his family had moved in about a decade ago, when they'd immigrated from Wiltshire, southwest England.

Though he was now permanently settled in Canada, Roger identified—with his crisp British accent—only as an Englishman, and every afternoon, he and his family of four engaged in a charming English ritual, for which I joined them: sipping tea in delicate floral Victorian cups, which they'd brought with them from England.

Roger, in his early fifties, had bright blue eyes, close-cropped gray hair, flushed cheeks, and economical lips in a prominent jaw. He was contagiously jolly, booming when he laughed. His wife, Julie, was a personal support worker in a nursing home, and her petite form and mild demeanor belied a high level of efficiency and energy.

Roger and Julie's son, James, was a quiet high school student skilled at carving chairs, canes, and photo frames out of wood. Their daughter, Emily, was a blond, blue-eyed middle school student with a talent for photography. In addition to sharing a dedication to tea, the Harleys shared a passion for the English sport of rugby. Roger had played in his youth, and both James and Emily played competitively at school.

"My father was a veterinary surgeon and I grew up around farm animals, with a connection to them and a respect for them," Roger told me. "Julie's also spent her life around animals. Her parents had a dairy farm. Julie and I had our own farm of sheep and cattle in England. We had forty-five hundred acres of land, on which we grazed up to a thousand cattle and twenty-five hundred sheep and lambs. But when mad cow disease and foot-and-mouth disease broke out in the U.K., farm animal prices plummeted and borders closed. Our farm never had a single case of mad cow or foot-and-mouth disease, but we suffered because others in the industry had it. We shut our farm down and moved to Canada. We started a farm here in 2002, called Harley Farms, where we raise pigs, sheep, and cows outdoors, as they're meant to be raised.

"Three years after we started Harley Farms, a company called Rowe Farms contacted us. They wanted us to join forces with them. Rowe Farms is a network of a dozen or so farms; our Harley Farms is the flagship. Rowe Farms sells eggs in lots of places, and beef, lamb, pork, chicken, and turkey mostly in specialty Rowe stores. It was founded in 1967 with the aim of raising farm animals without antibiotics and growth-promoting hormones. Since my joining, the focus has shifted broadly to animal welfare.

"Everything begins with animal welfare. Once you fix animal welfare, everything else is fine—you don't need antibiotics or anything else. It's not bacteria that's the problem in factory farms. It's that the animals are stressed and fed up. You may have noticed, my e-mail address has the term 'freedom foods' in it. That's because we believe that all animals should be raised with freedom. Animals deserve all the contentments in life.

"Our motto at Harley Farms is: 'where different is good.' When we first moved here, everyone thought that our ideas were bloody crazy and that we were absolute lunatics. But we showed them that our ideas can work, that there doesn't have to be a dichotomy between animal welfare and making a living. It's only in the last few years that people have started listening to us and taking our work seriously. People just can't believe what we're doing—they're amazed!"

After my rugged rooms in the Miller and Roberts homes, I was pleased to see that my room in the Harley home was comfortable, clean, and warm, with a cushiony bed.

I slept well, but not enough, for I was up before dawn to get a tour of the farm with Roger, an enthusiastic early riser if ever there was one. Despite my initial grumbles, I was outdoors when the sun rose, and I saw that the sunrise was magnificent, its molten orange and pink hues spreading like a stain on the fabric of the night sky, crisscrossed by wisps of clouds that resembled the inspired strokes of a painter's brush.

The headquarters of Harley Farms was a wood barn that was home to an elderly gray donkey, two young brown horses, two lively black and brown turkeys, a flock of hens, and a flock of elegant, polka-dotted guinea fowl with feathers as soft as silk. On winter nights, this diverse menagerie was joined by a herd of sheep.

Roger and I climbed into his tractor and he gave me a tour of his lands, which encompassed about 1,200 acres. The tractor was pursued by five energetic dogs.

"Our farm would be untenable without our five dogs," Roger said. Two of the dogs were herd dogs—little dark shepherds who helped herd cows and sheep. The three other dogs were guard dogs—enormous, snow-white, polar bear–like sentinels, there to protect cows and sheep from coyotes after dark.

The herd dogs formed a small part of a vast pastoral concept, which was: "Everything has to be mobile." Roger explained: "We rotate the animals to make sure the land's not being overused. So, for example, in any one part of the farm, we'll have two years of crops, then two years of hay, then a year of cows, a year of sheep, and a year of pigs. That's our farm cycle; it's seven years long. Different parts of our farm are in different parts of the cycle at any one time. Some parts of the farm have pigs right now, some have sheep, and some have cows. All animals play their own part in the farm cycle. Every farm has to have a cycle if it wants to be sustainable and maintain the land successfully year over year."

Roger's talk of natural farm cycles enchanted me because the factory farms I'd visited so far had been as devoid of cycles as cardboard. Every day and night had beheld the same misery, every season had glowed with the same darkness, every year had repeated the same tribulations. The only cycle had been suffering. Everything Roger was saying was so different from what I'd heard in the Robertses' community that he could just as well have been speaking a different language.

"Right now, we have a total of around three hundred sheep, pigs, and cows at our farm. We keep sows five to an acre, sheep three to an acre, and cows one to an acre."

An acre is about three quarters the size of an American football field—a considerable amount of space. In comparison to the factory farms I'd seen, this was a culture shock—giving animals so much room seemed indulgent. "Why don't you increase the number of animals per acre?" I asked Roger.

"Industrial farms grow by increasing the number of animals per acre—or per square foot, rather. We grow by increasing the number of acres. That's an important distinction between a pastoral farm like ours and an industrial farm. Industrial farms focus only on quantity; we focus on quality before quantity. We can't crowd animals together, because that reduces our animals' quality of life and it ruins our land. At our farm, it's all about balance."

This was real, thoughtful, ecological, ethical farming. It was exactly what I was looking for—*better* than what I was looking for.

Roger and I arrived at a herd of cows and descended from the tractor. The cows were petite, portly, and resplendent in appearance. Their breed was called Belted Galloways, and the cows were reminiscent of Oreo cookies, with a creamy white belt in the center and smooth black sides. They had two thick, ruffled coats of hair in order to not just survive but thrive in winters. Roger dropped two bales of hay in their midst using the prongs of his tractor. The cows approached and started sniffing and chewing through it.

Roger and I proceeded next to the sheep, who occupied their own areas of the farm. Of the Wiltshire Horn breed, they were white-haired, long-faced, and solemn and silent in appearance.

"I've selected my breed of cattle, pigs, and sheep carefully," Roger said. "The first piece of advice I always give farmers is: select the right breed. Breed is incredibly important for animal welfare. What I did here is test various breeds of cattle, pigs, and sheep to see which would work best in a humane agriculture. At one point, I had twenty breeds of cattle! I wanted to know which of them could be raised outdoors year-round—without getting cold, without being de-horned and castrated, *while* converting the food they eat effectively. The answer, I found, was Belted Galloways, a Scottish breed."

Roger was not just a farmer but an innovator.

He and I climbed into his tractor again and proceeded next to the sows. Pigs were Roger's favorite animals on the planet. He was their salesman, extolling their virtues to anyone who'd listen. "Pigs are so intelligent!" he boomed. "They're bold in their opinions, they're curious about everything, and they're stubborn. You can't get them to do anything they don't want to do! I respect that."

Roger's sows were no less unusual than his cows. Of the Tamworth breed, the sows were auburn in color, and they had deep-set eyes and twitching snouts. Their elevated back hooves made them look like they were tottering about in heels on the shimmering white carpet of snow. The sows lived in expansive pens, but from a distance it was impossible to tell that there was any limit to their roaming, for the boundary of their pens was no more than a single, thread-thin electric blue wire.

"The sows can just step over the wire and leave any time they want to," Roger said. "But they don't want to, because they're happy here. Our Tamworths are among the oldest English pig breeds and are a natural choice for outdoor farming. They have a good, thick covering of hair and two to three *inches* of back fat." Roger demonstrated the hefty amount with his thumb and forefinger. "Our pigs *enjoy* the winter; they *love* burrowing in the snow. In comparison, the usual industrial pink pigs have very little back fat, so you can't turn them out in the cold.

"Industrial pigs also have all sorts of health issues. They have very little disease resistance because of their genetics and their environment. Our pigs have great disease resistance. They're very active in their lifestyle, and they hardly ever get sick, touch wood. They're very protective, very good mothers. Industrial pink pigs, in comparison, are fantastic at getting fat, but they have no idea of how to look after their own piglets. The mothering instincts have been bred out of them completely."

I thought back to Charlie's pig farm. His sows had looked sick and comatose. I'd held a castrated, tail-docked piglet, and the piglet had squealed frightfully, but his antibiotic-laden mother hadn't even bothered to look up at him. The difference between Roger's and Charlie's sows could not be overstated.

Roger's sows drooled in anticipation as we gave them their daily breakfast of wheat, oats, barley, and corn, which Roger and I had loaded in pails into the back-wagon of the tractor earlier. Mixed grains were the sows' first meal of the day; their remaining day's rations consisted of hay and haylage (fermented hay), which were piled up in their pens in mountains—an eternal green buffet.

"Pigs are given only a corn-based diet in industrial farms because those farms want to automate everything," Roger explained to me. "They want to press one button and have the food there. But you can't do that with hay. You can't mix it in water and move it through overhead pipes like you do with corn. Hay takes more work, more time, it's harder to manage. It's also harder to measure. When animals eat hay or grass, you don't know exactly how much they're eating, or when they're eating it. Industrial farms don't like that because they want to *control* everything. They want to measure everything to the decimal point, put it all into a computer, and let the computer do the thinking for them. For our kind of farm, you need farmers with a whole new mindset."

Pasture-based farming, it seemed to me, is a little like dancing. It involves a high level of care but also an ability to shimmy to the music of life. It necessitates an understanding and an acceptance that everything cannot, and should not, be measured, monitored, and molded to human whim. It represents a paradigm shift.

Roger's sows lived in roomy wood houses with a sloping roof and white canvas door. As the sows ate, Roger encouraged me to step, or crawl, into one of the houses. I did, on all fours, at first hesitantly, but then enthusiastically, for the house was lovely. It was bedded with straw and was spotlessly clean and surprisingly warm. The breeze outside was cold and crisp, but inside, the temperature was a sauna in comparison. The sow houses were another of Roger's successful experiments.

"I designed the sow houses, and my son James and his friends built them," Roger said. "In total, we tried somewhere between forty and fifty different kinds of houses for the sows. We took lots of factors into account—building material, space, comfort, cost, mobility, and wind direction. We found that this sort of wood house worked best. The sows just love it. They spend their days walking around outside and their nights warm and snuggled up inside."

As we crawled out of the sow house, Roger became more serious. "I worked at an industrial sow farm once, thirty years ago, in my early twenties, in England," he said. "It was horrible. The sows were tethered all the time. I felt sick from all the ammonia and gases, and I always had the bloody smell on me, even after a shower. You wouldn't last more than three or four hours in such a place, I guarantee you."

Roger was right. My night at Charlie's pig farm had been the worst of my life. I wouldn't wish it on anyone.

"When sows are tethered all day," Roger continued, "they get so used to standing still and their legs become so weak that if you put them in a barn, they don't move. It's not because they don't *want* to move, it's because they *can't* move. It's like if you've been bedridden for a long time—you can't just get up and start moving all of a sudden. I broke all the bones in my hand once, and it took me three months to get the muscles working properly again. Sows are treated so badly in industrial systems."

Life for pigs has only become worse since Roger's time. Sows today are not tethered, but, worse, crated—serving life sentences in iron-galvanized steel boxes, prisoners of crimes they have not committed. Pig operations have also exploded in size. The average pig farm in the United States had ten pigs in 1935 and a hundred in 1980. Today, it has more than five thousand.

Roger and I got into his tractor and proceeded onward to a charming wood barn of seventy auburn piglets. Two to three months old, they lived on a lush bed of hay and straw, which they ate, too. Like their mothers, they kept their bed remarkably clean, cooperating to keep their bathroom activities along one wall.

I breathed deeply; the barn smelled . . . *good*. Pastoral farms smell good, I realized then, because soil and straw absorb manure and odor. The open concept construction also helps with circulation, taking out stale air and conveying in fresh air. Factory farms cannot help but stink because their inorganic substrates—concrete and steel—cannot absorb odor, and they have no natural substrates that can. In addition, they have little ventilation, their only windows to the outside world being grime-covered exhaust fans.

As Roger and I stood in their midst, the piglets scuttled in a squealing brown stream from one end of the barn to another. I'd assumed this was their normal (strange) behavior, until Roger said "It's because you're here."

He was right; they stared only at me as they ran, paying no heed to him, as accustomed to his presence as to the clouds in the sky. I became astonished by how attuned they were to me. Even when I moved just one toe—within just one boot, within the hay—they detected it and reacted to it by surging faster.

Eventually, after half an hour, they lost interest and resumed their usual scope of activities. They rooted in the hay. They played. They nuzzled one another. They chased one another. They crunched ice—their favorite pastime, Roger believed. They trotted up to me and started chewing my boots. I waved them away. They returned. I waved them away again. They returned again.

It was like a piglet playground. Are these happy pigs, I asked myself. Yes, without a doubt. Animal happiness is almost as easy to see as human happiness. Unhappy pigs, such as Charlie's, bite each other's tails until they bleed. Happy pigs, such as Roger's, are calm and contented, mischievous and adventurous.

Roger audited the other farms in the Rowe network annually to ensure they were continuing to meet humane standards.

Roger, an internal Rowe auditor, was often accompanied by an external auditor, a woman by the name of Maud. She was not a government auditor, as one would expect, but was rather employed by a corporation, Société Générale de Surveillance (SGS). In both the United States and Canada, the relevant government bodies—the United States Department of Agriculture and the Canadian Food Inspection Agency (CFIA)—delegate most of the responsibility of farm certification to corporations and nonprofit organizations.

Maud, the SGS auditor, was not paid by the government but by the farms she certified, in the form of their hefty fees to SGS—hundreds of dollars per year per farm. Such a system creates an obvious conflict of interest: the audited party, the farm, is essentially a client of the auditor. If the auditor does not certify the farm, the farm no longer remains a client, and the certifying company loses hundreds of dollars of revenue a year. Consequently, almost all farms retain their certifications year after year. Less than one-tenth of one percent of American organic farms have their certifications revoked in any one year.

Alex Avery, author of *The Truth About Organic Foods*, is critical of the audit and certification system of organic and other alternative farms. In the documentary *In Organic We Trust*, he explains: "The certifiers are in business for themselves. . . . It's a *serious* inherent conflict of interest. . . . I don't think there's been any operation that's lost its USDA organic certification status because somebody's blown the whistle. And I think it shows that the system is rather toothless."

Far from being at odds, the incentives of the auditor and farmer in agriculture are perfectly aligned, their fates joined like two sides of a coin. Alternative farmers themselves suffer from such a lackadaisical certification system, as it is exorbitantly costly and reduces their credibility. The costs get passed on to consumers, and are one of the reasons that alternative food costs more. To be more effective, the farm certification and audit system should be public instead of private, done by the government instead of corporations.

In the case of Rowe Farms, the quality of the audit did not matter, because Roger's internal audit was more thorough than any external audit could be anyway. But poor-quality audits are a problem for most farms because they don't have such strong, consistent internal standards in place.

I accompanied Roger on an audit of three Rowe beef cattle farms, my expectations soberingly in check after Roger warned: "Our auditor Maud hardly speaks a word of English."

At the first farm, Maud, who was supposed to have arrived, was nowhere to be seen. As we waited for her, Roger introduced me to the farmers: a kindly middle-aged couple who, unable to have children of their own, "treat their cattle like their children." They had a herd of eighty glossy brown, black, and golden cattle.

Maud eventually arrived, without apology, over half an hour late. Middle-aged, Croatian, she had a sallow face surrounded by parched hair and adorned with a frizzy girlish fringe, which resembled a crumpled awning over a window. She did not like me, a fact that she made clear immediately.

"Whatareyou doinghere?" she demanded. "Sorry?" "Whatareyou doinghere?" "I'm learning more about animal agriculture." "Notmyside," she replied. "What?" "*NOTMYSIDE.*"

Did she mean "not my style," I wondered, and what did that mean in the context?

Roger somehow understood Maud and acted as her interpreter, repeating her comments and commandments in his ringing English accent. "Howmanycats?" Maud demanded of the soft-spoken farm couple. "Sorry?" they said. "HOW MANY CATS DO YOU HAVE?" Roger boomed on her behalf. "Are cats part of the audit?"

they asked in confusion. The right answer would have been no, but Maud did not deign to reply. Roger explained to me later: "She has a thing for cats. She's always asking about them."

Maud asked the farm couple a few perfunctory questions about well water, appearing as bored by the questions as by their answers. She then walked briskly around the cattle barn instead of into it. Her review of its perimeter marked the end of the audit. Her inspection methodology was: Check nothing, approve everything.

"I haven't a clue as to why she walks around barns," Roger told me later. "Maybe she's looking for cats. She keeps audits short because she doesn't like being outside. But her audits aren't useful anyway, whether they're long or short. She uses my audit forms in writing up her own, and it would be easier if I could save everyone the trouble and mail them out to her, but the government requires us to have these audits. . . . Why don't we go with a better auditor, you're asking? Because others are just like her. They don't have a clue."

Roger and I proceeded to the second farm. Maud, who was supposed to be following us, disappeared along the way, and we had to wait for her again. The second farmer was a seventy-two-year-old boxing bag of a man, well over six feet in height and close to three hundred pounds in weight. His fingers were sausage-like, each of his hands equaling two of my own, with my foot thrown in too. He was a seventh-generation farmer, a man who today farmed not for the money, but for the pleasure of continuing the tradition his settler ancestors had commenced close to two centuries ago. His 700 acres of land were enjoyed by only thirty-five cows, leading Roger to exclaim, "His cows get ten to twenty acres each!"

When Maud eventually arrived, she peered at his barn cats, cast no more than a cursory glance at his cattle, inquired about well water as before, and marched in a circuit around the perimeter of the barn. We then left for the third and final Rowe farm for the day, where the audit was similarly vacuous.

Every evening at six, the four Harleys and I would sit down for a delicious dinner prepared by Julie. Afterward, the children would watch television, and Roger and I would shift to the sofa, where I would ask him questions about animal agriculture.

"What do you think of organic?" I asked him.

"Organic needs to be strengthened and tightened. A lot of organic is supplied by big companies whose lawyers are paid to figure out how to get within a hair's breadth of the law. There's also a lot of mislabeling in organic. Companies are selling meat, milk, and eggs as organic that aren't *actually* organic, or even close to organic, but they're not caught because of the certification and audit system. You've seen how the system is. Also, organic *should* mean that animals can range and roam freely, but it doesn't mean that right now. Animals are out only on certain days of the year, and even then, hardly. A lot of organic right now is about making the most money possible off the public while doing as little as possible. It's a scam."

Brick Roberts had also described free-range as a scam. "How does Rowe Farms market its products if not as organic?" I asked.

"We describe our work as 'quality with a conscience.' Our goal is to build a good brand that people trust. We used to call everything 'naturally raised,' but the government said we can't use that term anymore."

The Canadian government has taken a strangely philosophical stance on the term "naturally raised," stating that domestic animals can never be naturally raised, and only wild animals can. In the process of policy-setting, it has forgotten the definition of "wild" animals: animals who are *not* raised by humans. The term "naturally raised" has been hotly contested in the United States as well, which has erred in the opposite direction of its northern neighbor, choosing to define the term too widely. Even if animals are confined to dark, dingy cages, they can still be called "naturally raised" in the U.S.

"Use of a naturally raised marketing claim standard has the potential," explains the USDA in a news release, "to increase the available supply of U.S. meat products . . . in the United States, the European Union, and other export markets."

"What do you think of the idea that small farms are the solution to agriculture?"

"They're not," Roger said. "Scale of economy is everything. Small is beautiful in most respects except when it comes to the pocketbook. The question in agriculture today isn't whether there should be scale of economy—there *should* be scale of economy—but whether that scale should be supplied by a natural outdoor farm or a factory farm. In other words, we're not talking about size, we're talking about system. Our Harley Farms is growing a lot every year. We're somewhat small right now only because we're new. The farm we had in England had thousands of cows and sheep, all of whom had good, natural, happy lives."

"I see. That's interesting—differentiating between size and system. I hadn't thought about it that way before. On a different note, I've seen farms but I haven't seen slaughter. How's slaughter?"

"I've audited slaughter plants, like I've audited farms, and there's lots of problems in slaughter. There's too many people in the industry who turn a blind eye to everything. . . . I have a friend who's Pakistani, whose name's Abdul. One of his sons recently got a heart transplant—among the very few kids in the country to have gotten one. Anyhow, Abdul owns a slaughter plant. His brother owns a slaughter plant, too, and his brother's become a vegetarian! Can you believe it—a slaughter plant owner who's a vegetarian?"

"I can believe it. How can I visit a slaughter plant?"

"Trust me, you *don't* want to see slaughter. You won't like it. I don't like it. And it's impossible to get in. I can't see how you'd manage."

"Okay. On to something else: how do you think change can be brought to animal agriculture?"

"The people who're beginning to have the most power is the public. It's all starting to come down to consumers. But the problem is that consumers haven't got a bloody clue. They hate agriculture—they don't trust it—but they also have no idea of how animals *should* be farmed. Also, people have a complete and utter disconnect between farm animals and cats and dogs. They spend loads of money on their cats and dogs, but they don't want to pay a few extra cents for higher-welfare, pasture-raised meat, milk, and eggs. There's a lot of confusion in the public mind, I'd say."

The amount of money Americans and Canadians splurge on their beloved cats and dogs is touching but mind-boggling. According to the American Pet Products Association, Americans lavished a grand sum of $55 billion on their pets in 2013, a threefold expansion from two decades earlier. An American household with a dog spends close to $900 on veterinary care alone every year. More than two thirds of American households today have one or more animals, which translates to children being more likely to grow up with pets than with both parents. Since childhood, we develop a powerful connection with cats and dogs and no connection whatsoever with cows, pigs, and chickens.

While there seems to be no cap in sight on pet spending, household spending on food has been bottoming out for decades. According to the USDA, Americans spend only 10 percent of disposable personal income on food, counting both groceries and dining out. American expenditures on food as a share of consumer expenditures are a fraction of that of other countries, and less than half that of many European countries.

"How do you think pastoral farms like yours can become more prevalent?" I asked Roger.

"You have to show farmers how this kind of farming can be profitable. Money talks. The fact is: we *can* economically farm animals in an environment as they were naturally intended to live."

"If pastoral farming is so doable, why don't more farmers do it?"

"Because farming's not flexible. Once you create a particular system, you can't change it. Corporations pay for factory-farm buildings, but then farmers have to do exactly what they say for the rest of their lives—even if they hate it. Debts take a whole generation to pay off. Once, a farmer with a big sow factory farm visited us here. As he was leaving, he told me that he could never look his sows in the eye again. He never wanted to set foot in his own sow factory farm again! But there was nothing he could do about it."

"Do you see any other challenges to change as well?"

"Yes. Factory-farm corporations grow to huge sizes and have billions of dollars of money. They try to buy out pastoral farmers with their money. One of Canada's largest factory-farm corporations came here one day several years ago, just a year or two after we'd started Harley Farms. There were two men wearing suits and holding briefcases. They came here to offer me a quarter of a million dollars per year to stop working and to stop telling everyone that factory farming is bad. They said that if I close down my farm and stop expressing my views, they'll pay me a quarter of a million dollars every year.

"What did you say?"

"I turned them down."

Roger was admirable for his dedication to his humane vision. Almost anyone else would have taken the quarter-million-dollar sum and checked out. The incident demonstrated how stealthily and silently a vision of a better future can be crushed, a visionary removed, a good farm vanished—without a word to the contrary. It showed that the industry of agriculture is at war: factory-farm

Goliaths—with their suits, briefcases, and wads of cash—against pastoral farmers like Roger.

On the positive side, the fact that a multi-billion-dollar corporation *can* be threatened by a single individual highlights the strength of individuals. Each of us is endowed with the power of choice; we just have to decide to make the right choice.

I accompanied Roger on his farm rounds every morning, trying my best to maintain my balance on the back-wagon of the tractor, though I tumbled occasionally.

On my last morning, I overslept—which meant that I awoke at eight. Roger, who teased me for my late mornings, began the farm rounds himself that morning. But he burst in through the front door as Julie and I were sipping tea in the kitchen and Emily was on her way out the door for school. "A calf was just born!" he exclaimed. "Let's shift her and her mother to the barn for a day so they can take it easy."

Emily and I excitedly followed him out the door. Emily, who'd been farming since she was a toddler, knew exactly what to do. She loaded hay onto the back-wagon of the tractor—"to make the wagon more appealing to the calf"—and off we went.

The cows, however, were gone. The land stood empty. Coming closer, we saw that they were hiding in the neighboring cedar trees. Emily and I remained in the tractor as Roger went off to locate the newborn calf among the trees. He emerged soon with a small, beautiful, wet, ruffled animal in his arms, black and white, with big ears and big eyes. Still with umbilical cord, under an hour old, she was the youngest animal I'd ever seen, the only winter-born calf at the farm.

Roger strode toward Emily and me with the calf. Bellowing, her mother came in hot pursuit of him, right on his heels. I feared she would charge him; he had been "flattened" by a cow once. To be fair, from this cow's perspective, he was a thief, running off with her first child, the most important thing in the world to her, whom she'd carried in her womb for nine long months.

Cows are among the most protective mothers on the planet, Roger had told me. As he'd put it: "When you go near a calf to put something in its ear, and the cow knocks you off your feet, you know she's a good mother."

It is the cow's formidable maternal instinct for which she is revered in India. "Mother cow is in many ways better than the mother who gave us birth," Gandhi said about cows. ". . . Well, I say this not to disparage the mother who gives us birth, but in order to show you the substantial reasons for my worshipping the cow . . . I worship it and I shall defend its worship against the whole world."

Like a thief, Roger had a strategy for survival. Holding the calf in his arms, he bellowed back to the cow, his comical *moo*s meant to reassure her. When he reached the wagon of the tractor and deposited the calf on the hay, an astounding event occurred. Emily, half my age, shifted to the driver's seat and started driving. "The cow will follow the wagon because her calf's in it," Emily explained to me.

To my astonishment, the cow did follow the slow-moving wagon. She'd given birth just an hour ago, but—her body heaving like a ship on water, her hooves pounding snow like dust—such was her love that she pursued the wagon, her nose touching that of her calf, who in turn leaned toward her. When we reached the barn, Roger picked the calf up and deposited her in a stable. He rubbed handfuls of hay over her body to warm and comfort her, then he left the stable and her mother took over.

In all the cow-and-calf business, Emily had missed her school bus and her morning classes. To me, a dedicated nerd in my student years, the idea of skipping school, for any reason, was unthinkable. To the Harleys, however, dedicated farmers that they were, the idea of neglecting animals, for any reason, was unthinkable. "Emily and James are late for school about once a month," Roger told me. "If James hadn't already gone to school, he'd be here, too, helping out with the cow and calf. At our farm, the animals always come first."

For the first time, I understood the term "family farm." A family farm is not just owned or managed by a family, but is a place where every member of the family is involved, to the best of his or her ability.

"There's so much more going on at our farm in the spring than in the winter," Roger told me. "Come back to see us in the spring, and bring your family, too!"

I did.

In the spring, Roger's village, Keene, consisted of verdant slopes and luxurious hills, of lilting leaves and singing birds.

A year at a pastoral farm can be viewed from the perspective of a single day, I realized as soon as I arrived. Spring is morning, summer is afternoon, autumn is evening, and winter is night. The Harley animals had, in a sense, been asleep in winter, and now, in the dawn of spring, they were rising, bounding with energy. Winter had been the season of silence; spring was the season of offspring. Calves, piglets, and lambs formed a noisy but peaceful nursery on pastures of budding grass and gushing springs.

Sheep were the quietest and least interesting animals in winter, but they were the loudest and most interesting animals in the spring, maintaining a continuous chorus of high-pitched *baa*s with their lambs. Lambs were tiny, adorable, curly-haired creatures. They skipped along cheerily beside their mothers, congregating sometimes in playful groups. When lambs slept, often huddled up in pairs, adult sheep "babysat" them, one or two of them keeping an eye on the lambs as the other mothers grazed.

The highlight of my winter visit had been the birth of a calf; the highlight of my spring visit was the birth of a lamb. Both births illustrated the depth of maternal emotion and instinct in farm animals.

In the shadows of the barn, a sheep was in the throes of labor. She was licking a lamb, though, which led me to assume that she'd given birth to him, and was just about to give birth to his twin. This was not so, Roger told me, eyeing the lamb, then taking him away. "If she'd just given birth to this lamb, it would be wet," he explained. "But it's dry. That means it's not her lamb; she stole it from another sheep. Sheep like to adopt each other's lambs. They just *love* lambs." He returned the lamb to the rightful mother, who accepted him gratefully, licking his face.

In the meantime, the laboring sheep gave birth. She stood up and a lamb simply fell out of her body. The lamb was yellow in color, covered with ringlet curls, about the size of a newborn baby. Birth is a miracle of nature, a commencement to the mysterious cycle of life, and I was touched to witness it.

The new mother started tenderly licking her lamb's face and neck. Then, something wonderful and unexpected happened. All the other surrounding sheep, dozens of them, visited the newborn lamb in a large circle. They were greeting him into the world, welcoming him into the herd. I realized then that animals have their own rites and rituals, evolved over thousands of years, instinctive and timeless, enchanting and enigmatic.

In the evening, my parents, sister, and I had dinner with Roger and his family at the lakeside restaurant of the rustic inn where we were staying, ten minutes away. The dinner conversation held a pleasant rhythm, except in certain rocky moments.

Roger and Julie were at a loss for words when my mother suggested to them, as she regularly did to me: "Why don't you get an office job like everyone else? Wouldn't it be much easier and more high-paying than what you are doing?" My parents didn't know what to say when Roger stated, in his passionate, elevated tone: "Industrial agriculture's *got* to go. It's *so* inhumane, and it's also one of the biggest polluters out there. . . ."

Visiting the Harleys with my family offered me yet another angle on the term "family farm." A family farm is a place where you can actually *bring* your family. My parents enjoyed looking at the white sheep and springy lambs, the Oreo cows and fuzzy calves, the protective sows and playful piglets.

You can gauge a farm's compassion by your family's reaction—outsiders form the litmus test.

CHAPTER SEVEN

RED, WHITE, AND BOB VEAL

CONFINEMENT AGRICULTURE

"Eating veal—or not eating it, to be more accurate—is one thing many carnivores and vegetarians can agree on," states an article in the *Washington Post.*

"Veal stirs the emotion like few other foods," states the *Independent* in the U.K., "only foie gras can compete in the public's mind as a clear, open and shut case of animal cruelty."

It was because of the rock-bottom reputation of veal that I was surprised to read on the website of the Ontario Farm Animal Council (which later changed its name to Farm & Food Care):

"In Ontario, members of the Ontario Veal Association would be happy to show you their farms, and answer any further questions. Most people who visit a veal farm are pleasantly surprised by the conditions these calves are kept in."

Really?

I called the Ontario Veal Association, an industry support group that, in recent years, has been marketing veal like candy, asserting that it is beneficial not only for the palate but also for the economy. The phone was transferred through two people before eventually reaching someone willing to talk to me. "Why do you want to visit a veal farm?" she asked suspiciously.

"I became interested in veal when I volunteered at a dairy farm."

"I'll need references from the dairy farm."

"I don't have references."

Michael Miller would never give me a reference, I knew. But also, if references were needed, what was all that on the website about veal farmers being "happy to show you their farms"?

The call was useless. I decided to try to persuade the Ontario Veal Association in person. They were scheduled to have a table at Canada's Outdoor Farm Show.

I took a train and then a taxi to the farm show, which was in a town called Woodstock, two hours from Toronto and upstate New York. There were hundreds of exhibiting companies at the show. One of them was dairy farmer Michael Miller, to my disappointment. I decided to give his booth a wide berth, hoping against hope I wouldn't run into him.

One of the largest tents at the show was occupied by a poultry company called Clark, whose egg facility had recently installed "an eight tier Farmer Automatic cage system," which was "the second installation of this type in the world." The new eight-floor cage system allowed the confinement under one roof of 135,000 hens—ten times as many as Brick Roberts had. Brick's four-floor cages had risen high above my head to the ceiling; how tall would eight-level cages be, I wondered with a shudder.

The Ontario Veal Association had no more than a small, humble table at the farm show. "The Ontario Veal Association does a lot of things," explained the young blond woman at the table. "We do consumer outreach, support for veal farmers, and veal promotion. We have a website called Ontario Veal Appeal, where we tell people why they should eat veal, and we give them recipes."

She pushed into my hands several veal pamphlets and leaflets, including a Veal Fun Book—an eight-page children's booklet consisting of coloring, crosswords, puzzles, and games about veal. The first page of the booklet, a puzzle to color in, showed a calf standing on grass under a grinning sun.

"So, you want to work in veal?" she asked me.

She was assuming I wanted to work in agriculture because almost everybody at the farm show worked in agriculture. Come to think of it, even the train conductor who'd stamped my train ticket had assumed I worked in agriculture because she'd seen me reading a poultry magazine. "I'm scared of eating chicken," she'd confessed, "and I don't know what to eat."

"Yes, I want to work in veal," I replied confidently at the veal booth, in the tone of voicing a life dream. The lie gave me a guilty, gnawing feeling in the pit of my stomach.

"You can get some more information and meet some veal farmers at the veal seminar tonight. Here's the address."

The seminar, titled "Veal, Grains and Oilseeds," was in the basement of a seedy, truck-driver Super 8 motel near the farm show.

The seminar was set to run from seven to nine in the evening. I arrived at a quarter to seven, with an agenda to leave by seven, for three reasons.

First, I was nervous. This was a seminar of veal farmers—people who worked in the industry and who I imagined would be able to see right through me, like through a pane of glass. Also, I kept imagining dairy farmer Michael Miller walking into the Super 8 basement. He would gasp to see me there, point a long finger at me, and yell out for everyone to hear: "She does *not* want to be a veal farmer." A final source of my stress was that the last train home would be leaving soon, and if I missed it, I'd have to stay at this grimy Super 8.

I'd expected there to be a horde of farmers at the seminar, but only two men and two couples attended, in addition to the two presenters. I sat down next to one of the two individual men.

His hair, which had appeared from afar to be a streaked blonde, was actually a streaked gray. He was older than I'd thought. "I started with one veal calf twenty-five years ago," he described happily, "and I now have three hundred of them."

"That sounds like a lot of calves," I said stupidly.

"It's not a lot compared to other farms," he said, laughing.

My intuition told me I would get no further with him, and since time was of the essence, I rose rudely and approached one of the two couples in attendance.

Their names were Pierce and Mary, and they were in their late twenties. Pierce was attractive, with a lean frame, sea-blue eyes, curly chestnut-brown hair, and a disarming smile. Mary had green eyes, ear-length blond hair, and was pregnant. She looked older than her husband, and did not possess his easy charm and good looks. Pierce was a full-time electrician and Mary a full-time nurse. In the early mornings and evenings, they tended to 200 veal calves.

I did not sit next to Pierce and Mary, but in front of them, because I thought it would appear more discreet. "Can I visit your veal farm?" I asked them urgently. My behavior reminded me of drug dealers in movies: making deals without making eye contact—all in a moment, under the table. There were only five minutes left to seven o'clock; time was running out.

"How come?" Pierce asked the back of my head.

"I'm thinking of starting a veal farm." I felt bad as soon as I uttered the lie, for Pierce seemed genuinely warm and helpful.

"Both of us work during the week," he said, "but you can drop by over the weekend."

I noted down Pierce and Mary's phone number and address, and rushed out just before the seminar began. I would be entering one of the most secretive places on the planet.

I decided to get Pierce and Mary a gift to show my gratitude. A baby present was appropriate, as they were expecting. A baby present relating to cows or calves seemed ideal, as they had a veal farm. I visited the children's section of the bookstore, and, to my amazement, found several books on cows and calves.

One of them posed a question as its title: *What Do Cows Do?* The text of the book answered, "Cows live on farms. Cows eat lots of green grass. . . . Cows have babies called calves. The calf drinks milk from its mom."

But dairy calves do not drink milk from their moms. They do not drink milk at all, usually, as farms profit more by selling milk to consumers than by giving it to calves, for whom it is intended by nature. With their gargantuan udders, Holstein dairy cows today are genetically bred to produce astounding quantities of milk—twenty times as much milk as their calves can actually drink—but dairy farms sell all the milk, keeping none of it for the calves.

Calves are instead given a watery milk replacement formula constituted of milk by-products. This practice is conceptually fitting for veal calves, as they are themselves a by-product of the dairy industry. In fact, the dairy industry views the veal industry as an important customer, not just because the latter purchases calves, but also because it purchases milk by-products to feed the calves. As such, it is a double customer, buying not just the car, but also the gas.

Moreover, *What Do Cows Do?* suggested that dairy calves stay with their mothers—a false statement. Dairy calves are separated from their mothers as soon as they emerge from the womb. Female calves are confined to hutches at the dairy farm itself, as I saw at the Miller farm, and they later replace their mothers as dairy cows. Male calves are considered worthless because they cannot give milk; they become veal.

"Dairy cows must give birth to continue producing milk," explains the U.S. Department of Agriculture, "but male dairy calves are of little or no value to the dairy farmer."

The dozen other children's books that I picked up about animal farms were also misrepresentative and outdated, and I could not bring myself to purchase any of them. What we learn as children—the images we see, the impressions we retain in our memory—inevitably become a foundation of our lifelong psyche. As adults, we do not throw up the artificial mental meals we are fed as kids; we keep chewing on them throughout our lives, and we defend them vociferously when faced with the truth. That's why it's important for our first meals to be true to begin with.

I got Pierce and Mary a small calf toy along with a set of baby towels.

I took a three-hour train ride to Pierce and Mary's rural community. I assumed I'd be able to hail or call a cab after getting off the train.

I was wrong. There were no cabs.

Desperate, I walked into a restaurant and asked the waitress: "Can your delivery person give me a ride?"

"We don't have a delivery person. We don't make deliveries. But if you want," she continued helpfully, "you can ask the guys at that table there if they'll give you a ride. They've been sitting here for a while and I don't think they have any plans for the day."

I followed her gaze to a table of four men. They were elderly—ranging in age from seventy to eighty—and obese—tipping the scales at 200 to 300 pounds. They had heavy, sagging faces, and lined, leathery necks, and they wore baggy jeans and stained sweatshirts. They did not inspire trust. Swaying on the spot, I thought of taking the next train home.

Eventually, I decided against it, for I'd worked hard to get here, and I could not concede defeat now. Pulling myself together, straightening my shoulders, I approached the men.

"I'm looking to go to a veal farm," I said. "Would one of you mind giving me a ride?"

None of them looked up at me. I realized that I'd been so nervous I was whispering. Clearing my throat, I voiced my request louder.

One of the men glanced up at me from his bacon and eggs. "CAN ANYONE GIVE THIS LOVELY YOUNG LADY A RIDE?" he boomed at the others. Their heads bounced up from their plates of eggs and bacon in startlement, as turtle heads emerge from their shells.

"I'll take ya after I finish my coffee," one of the men pronounced quietly.

His face was made of folds that looked like crumpled, overused tissue paper. His nose was hooked and curved like a vulture's, and his cheeks were ruddy, with a patchwork of veins running through them like underground rivulets. His moustache was bleach-white, and his teeth were yellow and tarnished—where there were teeth.

Several scattered spaces in his mouth were dark and sparse like deserts. His double chin was larger than his chin, and it tugged his entire face downward with it, exerting gravitational pull even on his baseball cap, which stated: WE SUPPORT OUR TROOPS.

I stood beside him awkwardly as he finished his coffee. He did not invite me to take a seat, nor did he introduce himself. When he finally rose, he did not bid good-bye to his breakfast-mates, but merely forced out a smile, which looked like a grimace. He stumbled away from the table, and I saw then that his belly affected the architecture of his entire person, making his legs look like small, servile sticks struggling to prop up a monument on the verge of collapse.

He began walking, but the direction of his stride was not the front door. He stomped toward the back door. He had not parked his car in front of the restaurant, but *behind* it, I realized with a dawning sense of alarm. As I followed him through the narrow, shadowy back corridors, the discovery threw me off entirely, for I wondered: What sort of person parks his car *behind* a restaurant? Probably a criminal—someone shirking the law. Maybe even a murderer or a rapist.

More important, what kind of person takes a ride from a potential criminal? *A fool.*

The silent, nameless man opened the final door, and then there we were, under the shining sun, and he was fumbling in the pocket of his jeans for his car keys, as I was fumbling in my mind for an excuse to leave. He found his keys before I found an excuse. I got in the car.

His name was Jim, and his car was small and stick shift, sprinkled with cigarette ashes. Placards on his back seat, matching the message of his baseball cap, stated: WE SUPPORT OUR TROOPS.

Jim drove like a maniac, stopping, starting, and swerving suddenly, like he was racing an invisible driver. He was a chain-smoker, and found smoking most flavorful with his windows closed. When I started coughing, he reluctantly opened his window a crack.

Jim had seemed quiet in the restaurant, but he was actually as talkative as a child, speaking to me in his hoarse smoker's voice in a way that suggested we'd known each other for years. "People don't help each other anymore," he lamented. "If yer sick, no one will come and check on ya. Yer all by yerself. No one knows their neighbors. . . . I don't know if ya've noticed, dear, but we live in a Communist country. . . . Things around the world don't look too promisin', dear. So few people have too much and too many people have too little. . . ."

Jim's appearance, tobacco addiction, and style and subject of speech reminded me favorably of Brick Roberts. The comparison made me feel safe, except when Jim continued, "Ya know, ya took a big risk by coming alone out here to the country without a car. Kidnappin's and rapin's happen all the time. There's a lotta screwballs around. And a pretty girl like ya. . . .

"I used to be two hundred and seventy-five pounds before I started eatin' cranberries with my eggs and bacon in the mornin'," Jim went on. "I lost forty pounds because of the cranberries, but I still have diabetes."

Jim was five-foot-eight in height and seventy-five years in age. At this stage in his life, he suffered from obesity and diabetes, but earlier in his life, in his childhood years, he'd suffered from hunger. Born in 1936, he'd dropped out of school in the third grade, during World War II, because he and his sister had had little to eat, and he'd had to find work. His life was hand to mouth until about a decade later, when he became a truck driver and drove more than three million miles across North America.

But he was forced to retire at forty-six when, under the pressure of loading heavy equipment onto a truck, his back "gave out" a second time. He passed his retirement with his sister and his niece until, about a decade ago, in the same year, both women in his life died.

"My sister was married to a guy who used to beat her. This one time after he beat her, I broke a couple of his ribs, and I took my sister and niece away. They lived with me after that. I provided for them. I treated my niece like a daughter—ya remind me of her, dear."

Perhaps that was why Jim had agreed to drive me. He was kind, genuine, and starved for human companionship. I felt guilty for having judged him earlier based on no more than his appearance and car-parking preferences.

"One day, when my niece was crossin' the street," Jim continued, "someone ran her over. She passed away. My sister also had a car accident. She didn't pass away immediately, but a couple years after. Her mind became like a baby's, and I had to put her in a nursin' home. The home was expensive, but when I wasn't there, they wouldn't take care of her. I went there every day to make sure they took care of her."

Jim's tragic story about his sister led me to an epiphany. It's important to let people enter animal farms for the same reason that it's important to let them enter nursing homes, asylums, schools, and any other places whose residents cannot adequately speak for themselves. Public visits act as a kind of audit.

The outdoor farm show I'd just visited was a perfect example. The dozens of cows who'd been brought to the show had been healthy and immaculately clean. Although I hadn't given them much thought at the time, I saw now that the reason for their state was that thousands of people attend the show every year. If the cows were to look unclean or unwell, at least some of the attendees would complain and perhaps boycott the show. Public visits do make a difference to how farm animals are treated.

After taking a few wrong turns, Jim and I eventually reached the veal farm. "I'll wait for ya in the car, dear," Jim offered.

Pierce approached as I stepped out of Jim's car. He seemed confused to see Jim, who was by now shrouded in a cloud of smoke. He also didn't look happy to see me.

"I'm sorry to ask you this," Pierce said, sounding embarrassed, "but you don't belong to any animal rights groups, do you?"

"No."

Pierce relaxed. "Just thought I'd check. Sometimes people from animal rights groups like PETA come to farms undercover. We don't want any undercover people around here."

With that, Pierce began his farm tour briskly, trailed by me, his black Labrador, and his wife, Mary. Pierce and Mary farmed more than a hundred acres of land, much of it planted with corn, soy, and wheat. The crop fields formed a graceful ring around the house.

Pierce was of Dutch descent, his parents having immigrated to Canada from the Netherlands decades ago. Like many other Dutch immigrants (including the Millers), his parents started a dairy farm. But in the early 1990s, in the wake of mad cow disease, prices of veal calves plummeted to levels as low as ten dollars per animal, and Pierce's parents spied a business opportunity. Like investors seeking bankrupt stocks, they bought some veal calves and switched from dairy to veal production. Pierce later purchased his parents' veal farm.

As a veal farmer, Pierce today obtained his supply of veal calves from four neighboring dairy farmers, one of whom was his sister. When a dairy cow would give birth to a male calf, the dairy farmers would call Pierce, and he would then drive over in his truck and return to his veal farm with the calf in the back.

His veal farm consisted of three stages, he explained to me, based on a calf's age.

The first stage, from the first to the seventh week of a calf's life, was hutches. Fourteen calves lived in these individual white enclosures, tethered by metal chains that dangled down from the roof like nooses. A small window was cut in the roof, and through it poured in sunshine, forming a radiant patch in each hutch. In those patches lay curled all the calves, attracted to the light like moths to a flame. The day was sunny and cheery, but the calves could not skip or explore—or even walk a step.

"Since calves are social animals, they should be kept in social groups wherever possible," states the European Food Safety Authority. "Tethering always causes problems for calves. . . . Individually housed calves should not be tethered. . . . Exercise is needed for normal bone and muscle development."

As I squatted in front of the hutches, the calves rose to their feet, wide-eyed and knobby-kneed. All of them were black and white, as expected, except for the two smallest, youngest ones, who were red and white. They were red, not black, Holsteins. The color red is a function of a recessive gene in Holsteins, whereas the color black is dominant.

"They're twins," Pierce said.

Myself a twin, I empathized with them sharply. Though the two brothers were born together, they were strangers, unable to touch or to play, and they looked the most forlorn of all their brethren.

The second stage of the veal farm was much better than the first, I was pleased to note. Fifteen calves lived together in a large pen, walking, nuzzling, playing, and lying. They stopped to stare at me briefly, then continued with their activities. "When we take calves out of the hutches and put them together in this barn, it's like bringing four-year-olds together," Pierce said with a laugh. "They get so excited! They run around and play."

The last stage of the veal farm was the coveralls, which was a simple, semi-circular, tent-like structure hewn of white fabric. Open from all four sides, it was pleasant and sunlit, consisting of eight spacious, straw-bedded group pens organized by calf age. Some calves in the coveralls were eating, some ambling about, and many were lying together at the back of their pens, glancing at the cornfields outside.

"They like being by the window because of the breeze," Mary told me, following the direction of my gaze. "We have two hundred veal calves in total. That's a lot less than many other veal farms. Some veal farms we know have six hundred to a thousand calves. A few even have two thousand calves. Those farms usually work on contract for companies, and they don't care too much about their calves. We do."

There was not a whiff of odor—the calves smelled better than I did. "I like keeping the calves clean," Pierce explained. "If they're not clean, they're not comfortable. I give them plenty of space and I change their straw often."

The Ontario Farm Animal Council stated, "Most people who visit a veal farm are pleasantly surprised by the conditions these calves are kept in." Given the industry's secrecy and its rock-bottom reputation, I *was* pleasantly surprised. How could it be that this was a veal farm? Yes, the calves were tethered for their first seven weeks, but the remaining time of their seven months was passed well. The visit was a wake-up call.

Earlier in the day, getting on the train, I would have asserted that a "*good* veal farm" is an oxymoron, just as I would have asserted prior to my stay with the dairy-farming Millers that a "*bad* organic farm" is an oxymoron. This was an education—it was a *good* veal farm. Pierce and Mary clearly cared about their animals.

The Ontario Veal Association was right to state, "Most do not understand veal production and most do not consider veal production methods in comparison to other livestock production practices." The comparison with other "livestock production practices" is the killer. These practices include, as I had seen, chained dairy cows, caged egg-laying hens, steel-crated sows, and grotesquely fast-growing chickens and turkeys confined to ammonia-swamped sheds.

As the star-spangled American flag has three colors—red, white, and blue—veal calves can be categorized into three kinds: red, white, and bob.

Red veal calves, like Pierce's and Mary's, tend to be housed in group pens and fed corn, for which reason they're also called "grain-fed" calves.

White veal calves, in contrast, are generally housed in individual crates or stalls and are fed only milk replacer, for which reason they're also called "milk-fed" calves. Their formula-milk diet, often kept deliberately low in iron, combines with their restriction of movement to atrophy their muscles and keep their flesh tender in texture, mild in flavor, and pale in color, hence the term "white" veal.

"White is not right," a BBC article states simply. Many farmers agree.

"White veal calves have health problems because they're not eating solid food," Mary told me. "By six to eight weeks of age, a calf naturally starts eating solid food—grass or grain—but at white veal farms, they'll keep feeding it only milk replacer. White veal farms are also fully closed up. There isn't any fresh air or sunlight like we have here. We don't support white veal."

Michael Miller, too, for all the problems at his dairy farm, was opposed to white veal, I remembered. "Calves shouldn't be kept in crates," he'd told me. "I sell my male calves to a red veal farm, never a white veal farm."

Because of their diet and living conditions, white veal calves are often ill. They're susceptible to *Salmonella* and *E. coli*, and to diarrhea and pneumonia. They often suffer from "iron deficiency leading to anemia" and "a lack of rumen development," according to the American Veterinary Medical Association.

In 2011, the Canadian veal industry experienced an outbreak of *Salmonella* Dublin, a dangerous antibiotic-resistant bacteria with a 50 percent mortality rate in calves, transmissible and deadly also to humans. The Ontario Veal Association itself blamed the living conditions of veal calves for the disease, stating, "Stress—from overcrowding, poor air quality, co-infections, transportation, or nutritional deficiencies—can trigger the symptoms of this bacterial infection."

As for the third kind of veal, bob veal, it refers primarily to age, the word "bob" relating to the word "baby." These male dairy calves are killed as soon as they're born, usually within just a day or two of birth. After spending nine months in their mother's womb, they spend no time at all on earth, their lives ending before they can begin—a tragic fate.

Bob veal created headlines in 2009. Undercover video footage from Bushway Packing, a slaughter plant in Vermont, revealed workers dragging, kicking, and electrically prodding bob veal calves. Some calves were "downers"—unable to stand up—because they'd been trucked long distances and were injured, starved, and dehydrated. Many calves were stunned carelessly at slaughter and were consequently alert and awake as they were shackled, butchered, and skinned.

Not only are bob veal calves killed upon birth, but the Bushway Packing case proved that their deaths can be particularly gruesome.

The animal agriculture industries of the United States and Canada—whether eggs, dairy, or pigs—tend to be similar, with the notable exception of veal. Canada produces red and white veal, and the United States produces white and bob veal. Red veal was developed in Canada primarily as an ethical alternative to white veal, gaining significant market share only in the early 1990s. Historically, white veal was "practically the only veal available to the consumer" in Canada, as in the United States.

The lives of American veal calves are thankfully improving over time. In 2007, the board of directors of the American Veal Association (AVA) voted unanimously to adopt a resolution calling for all American veal farms to transition from individual crates to group pens by 2017. In 2007, 90 to 95 percent of American white veal calves were crated or chained. By 2012, 70 percent of calves were housed in groups (assuming honest and accurate reportage on AVA's part). The national American shift from barren veal crates to group housing has also been aided by state legislation. Veal crates are today banned in Arizona, Colorado, Maine, Michigan, Rhode Island, and California.

Though about 30 percent of veal calves in the United States and about 20 percent of veal calves in Canada continue to be housed in crates today, none are in Europe. Veal crates have been banned in the United Kingdom since 1990, in Finland since 1996, and throughout the European Union since 2007.

Production trends in North America and Europe suggest that veal crates are on their way out of the world's farms, and may perhaps disappear entirely from the face of the earth in the span of a decade or a generation. Veal consumption trends, in parallel, suggest that veal is on its way out from the dinner plate.

The word "veal" originates from the Latin word *vitellus*, meaning young calf. Veal's roots lie in Italy and France, where it was associated with wealth and status. In Florence, it was once decreed that if veal was served at a wedding dinner, no other meat could appear on the same menu, on account of veal's supposed delectability. In the present day, the tables have turned on veal. In the 1991 comedy *Father of the Bride*, wedding planner Franck suggests veal for the menu. The bride, Annie, protests: "I have a problem with that. . . . I keep reading there's a lot of inhumane treatment in the way they treat the calves." Franck replies "I read that, too. Very chic."

In the 1980s, photos of suffering, crated veal calves sparked a bottomless drop in veal consumption; 1988 marked the last year that veal consumption in the United States stood at over one pound per person. Today, it sits at one third of a pound.

Thousands of people in the U.S. and U.K. took to the streets in the 1980s and 1990s to hold up placards denouncing veal, to stage boycotts, and to blockade trucks, ships, and airplanes carrying veal calves. Protesters even died for veal calves. In 1995, a thirty-one-year-old English mother, Jill Phipps, was crushed to death by a truck transporting veal calves, her spine broken under the wheels.

A 2013 *Rolling Stone* article describes consumers as having "voted with their feet, walking away in droves from veal raised in boxes about the size of a small child's coffin." The article continues: "There are, it seems, some limits on our taste for torture."

Pierce and Mary's farm tour took close to an hour, and I was afraid that Jim, who'd been waiting for me in his car, might have left.

He hadn't. I asked him if he could drop me off at the train station. He refused.

"Yer train's not for three hours," he said. "I'll give ya a tour of the country in my car until then." Though I conceded only reluctantly, the tour was, in retrospect, enjoyable and illuminating, both about agriculture and Jim.

We drove past a handful of cow herds grazing on lush fields, splashing green landscapes with their black and brown hues, their tails swaying merrily in the breeze. For every outdoor herd we passed, though, we saw several factory farms. If not for the metal feed bins standing beside them like sentries, they looked like anonymous warehouses.

Farms in the past were located in the country because there was room in the country for animals to roam. Today, the only reason farms are still located in the country is the decreased cost of country space, not the increased space per animal. If land were not cheaper in rural areas, factory farms could be located anywhere at all—in slums, suburbs, the North Pole.

"Poor creatures," Jim would say, sighing, when we'd drive past an animal factory. He would try to guess the kind of animal confined within: dairy cows, pigs, turkeys, chickens, egg-laying hens? When we passed one that he believed to confine egg-laying hens, his ruddy cheeks drooped with gloominess. "Farmers cut the hens' beaks in machines so they don't peck each other's eyes out," he lamented. "Farm animals today don't go out for a day. They go out for the first time only when they're taken to slaughter. It's all confinement agriculture now."

Jim's term "confinement agriculture" was exactly right, connoting a sense of imprisonment and punishment. The agriculture industry seems to view confinement as not just a means to an end, but an end in itself, undertaking the task with zeal.

"There's too many people who have twice as many animals on their farms as they should," Jim continued. "Animals get very crowded. It's like thirty people in one little room—that's how crowded it gets. I'm just glad to see a young person like ya interested in farming. But I hope ya don't keep too many animals on yer farm, dear."

I turned my face toward the window so Jim couldn't see my guilt-ridden expression. I'd told him what I'd told Pierce and Mary, that my dream in life was to be a veal farmer. "I won't overcrowd my farm," I promised him. The conversation was ironic.

Jim had been born in the country, and he'd lived his whole life in the country. I, in contrast, had been born in a city, and I'd lived most of my life in cities. And yet our views on animal treatment and "confinement agriculture" were the same. Both of us believed that farm animals should not be made to suffer and that we owe them better. No so-called rural-urban divide separated us. My investigations into agriculture were showing me that in addition to the rural-urban divide, *most* societal divides are overstated, age, race, gender, and generation among them. We're all more similar than we are different.

The tour of Jim's life that day was tender but tragic, and in that respect, it paralleled the tour of the country. I was happy that at the very least, I was able to repay Jim's ride with the company he longed for.

My favorite part of the day was when Jim drove us to a beach. The breeze was salt-filled and calming, the gush of water serene and soothing. We sat at a picnic bench, and Jim ate a hot dog he bought at a roadside stand, as I ate a sandwich I'd brought along with me. We looked upon the ocean and admired the white and gray seagulls flitting above the waves like kites. How gloriously free they seemed.

CHAPTER EIGHT

SLAUGHTER HORROR

UNRELIEVED AND NEEDLESS SUFFERING

I entered two words into Google: "slaughter plant."

One result, Blackwater, came up with a phone number and address. It was located an hour and a half northeast of Toronto and less than three hours from upstate New York. I dialed the phone number, hoping no one would pick up.

"Hello? Nader here."

"Hi, Nader. . . . Hmm . . . I'd like to visit Blackwater."

He burst out laughing. "You make it sound so *polite*, like you're going to a zoo or something! Why do you wanna visit? You wanna work here?"

This was the same assumption the Ontario Veal Association had made—that I was looking to work in the industry. "Yes," I replied gamely. "How do I get to you by public transport?"

This set Nader howling again. I seemed to be the funniest person he'd ever spoken to. "Are you kidding me? There's no trains or buses out here!"

But then he thought about it out loud and constructed a complex itinerary for me. "Take the subway, then a train, then a bus. . . . Tell the bus driver to let you out at a blue van next to a red mailbox. Got it? *Blue* van. *Red* mailbox. One of my guys will be waiting for you in the blue van."

This sounded like a *Sopranos*-style murder set-up. Knowing I'd lose my nerve if I didn't go to Blackwater immediately, I told Nader I'd be there the next day.

"You want to help me kill?" he asked. He was not laughing this time.

"Hmm . . . okay."

"Just kidding!" he laughed. "Tomorrow's your first day, so you can just watch. I'll let you kill next time you come."

We hung up, but half an hour later, Nader called me again. "Wear exercise clothes," he instructed. "I'll give you boots, a coat, and a hairnet when you get here." Half an hour later, he called me yet again. "Wear two pairs of socks, okay? It's going to be freezing." Half an hour later, he called me a third time. "Don't bring no lunch," he ordered. "We'll have lunch here."

"I'd really prefer to bring my own lunch."

"Don't bring no lunch, I said."

What would I eat at Blackwater? Meat? I'd been vegetarian for years, and I had a feeling I'd want to eat animals less than ever after seeing slaughter. Another issue was Nader—I had a feeling something was very wrong with him.

Consumed with numbing dread and slow-setting panic, I tossed and turned in bed for hours that night. But a mere hour or two after I'd fallen asleep, the phone rang. It was a quarter past five in the morning.

It was Nader. He hung up without saying a word. Fifteen minutes later, he called me again. "I see you do that *ALL* the time!" he screamed. He was yelling at someone near him, not me, but his voice chilled me. He hung up again without a word to me. Fifteen minutes later, he called me yet again, and asked, "Are you on your way? Are you on your way?" When I reluctantly got on my way, he called me twice again. "Are you on your way? Are you on your way?"

I felt sure that Nader had psychological issues, and the idea of entering a blue van beside a red mailbox started to seem as safe as stepping into the mouth of a python. On my convoluted journey to Blackwater, exhausting all forms of public transport, my dread magnified and swallowed my empty stomach into its vortex until I keeled over with nausea. But then there I was, stepping off the bus and entering a blue van parked next to a red mailbox.

The man behind the wheel was called Moss and he was a mountain of a man. His top four teeth were missing, and his mouth was like a deforested, gaping black hole. He probably lost his teeth in a fight, I imagined. Was he a thug? As subtly as possible, I asked him a series of questions to determine the answer.

Moss said he was born in Saskatoon, Canada, to Ukrainian immigrants. Today, in his late fifties, he was a divorced father to a son and daughter in their thirties. In his twenties, he'd been a social worker, working with drug addicts and homeless people, and then he'd become a miner, toiling under the surface of the earth in temperatures that often plummeted to levels far below freezing. His present profession, his final profession, was slaughter work.

Moss had gentle eyes that crinkled at the edges like a fire, and he had a low smoker's voice that I'd strangely begun to find comforting after all my farm trips. I decided it didn't matter whether he'd once been a thug; I liked him.

I hoped the drive to the slaughter plant would be long, but it lasted only fifteen minutes. Blackwater, from the outside, looked small, square, and concrete, resembling a dentist's office. Even its lobby looked like a dentist's office, with a white board and welcome sign. The pungent smell, however, gave it away—it was heavy with blood. Even the white floor was stained with tangling, twirling streaks of blood.

Nader greeted me in the lobby.

He was tall, brown-skinned, and muscular. His face was shaped like an uneven clump of dough and dented with teenage acne scars. His eyes were hidden behind glasses and his lips were thin and firm. He wore a bright yellow hard hat like a construction worker's and a white robe that resembled a lab coat. Over the robe, he wore a bright blue armor-like apron, and on top of that, a yellow shield-like apron. He appeared dressed for war.

We shook hands like we were about to enter into a business partnership. His hand was wet; he must have washed it a moment ago to shake mine.

Nader handed me dark blue scrubs, a hairnet, knee-high black boots, and a white robe. "Robes are for inspectors, but we wear them over our scrubs in winter because of the cold," he said.

I wore the scrubs and white robe over my own layers of clothes. But both were extra-large men's size and far too loose for me. I feared they would fall off, as a shell around a peanut.

The boots, also, were too roomy for my feet, the same size as Nader's boots.

After changing, I wavered at the steel door to the kill floor, trembling at the wailing churn and heavy thud of killing machinery seeping through the cracks. There was a memory I'd been trying to banish from my mind, but it flooded in as I stood there. As an eighteen-year-old, I'd taken a wounded stray cat to the veterinarian. As he'd prodded and examined the wound, the cat had screeched and screamed. I'd looked at the bleeding pink flesh and a lightbulb had extinguished in my mind. I'd fainted.

Nader believed I wanted to *work* in slaughter, but I was afraid I might *faint* at Blackwater.

As I pushed open the door to the kill floor, I knew I would not be walking out the same person I was walking in.

The kill floor was about the size of a an apartment. The walls were devoid of windows, and the colors of the floor were gray and red: gray machines, meant to meticulously convert life to death, and a stinking, decomposing, oxidizing river of red on the floor.

In order to reach Nader on the other side of the kill floor, his back turned to me, I had to step tremulously into the pool of blood and body fragments. I put forth one foot, then the other, feeling repulsed, trying my best to avoid splashes, the challenge of traversing the red seeming as life-threatening as crossing quicksand.

My foot slipped. I almost fell. Blood is more slippery than ice, I learned then—my first crucial lesson of the kill floor.

When I reached Nader, he retreated into a room off the kill floor, called the holding room, where animals are unloaded from trucks. I followed him there. The holding room's seven pens held around forty sheep and ten goats (they'd held ten more sheep and goats before I'd arrived). Some holding pens were bursting at the seams, with sheep and goats crammed together body to body, head to head, like cans of beans. Still and calm nonetheless, they seemed ignorant of the fate that awaited them in a matter of minutes or hours. I noticed that they had no food or water, even though water is mandated by law at all times (and food is mandated if animals are held for more than twenty-four hours).

In the holding room, Nader lit a cigarette and gave me a tour of his life. "I'm Muslim," he said. "My parents are ethnically Indian, but I was born in Guyana. My family moved from Guyana to Canada when I was ten. I had the typical West Indian dad who beat me, so I ran away from home when I was twelve. I needed money, so I dropped out of school when I was sixteen. I started working in slaughter, and that's what I've been doing since then. I'm thirty-two now . . . I can't read or write good, but why do you need it? . . . I've been working in slaughter for half my life, but I've never had a single day of slaughter training."

Slaughter worker training is strictly mandated by law in Canada, and it was a testament to the looseness of the system that Nader had been slaughtering for half his life without a day of training. Studies find that insufficient worker training is a contributing factor to inhumane animal treatment at slaughter.

"But when I work a full day here," Nader continued, "I make a hundred and forty dollars *clean*."

"What does 'clean' mean?" I asked him.

"It means under the table."

"What does 'under the table' mean?"

"It means nobody gotta know. It means I'm not in the system, I don't get found out."

Nader's eyes settled on the youngest animal in the holding room, a curly-haired, wide-eyed white lamb. "All the other sheep and goats we're killing today are old females," he told me. "They're called 'nanny goats' or 'nanny sheep,' I guess. We're not going to make much money off them because their meat isn't tender. We're going to sell them cheap. White people don't eat that kind of old meat; it's all going to ethnic stores and restaurants—Jamaican, Haitian, Guyanese. But that thing there, that lamb, will be the most tender. It'll get us the highest price."

Nader's cigarette ended and, with it, my reprieve in the holding room.

Nader reached for a sheep and dragged her out of her pen by a hind leg. *Baaa! Baaaa!*

He dragged her onto the kill floor and closed the holding room door behind him. He relinquished her leg and picked up a stun gun. As the sheep jumped to her feet, he held the gun to her forehead and pulled the trigger. The sheep thudded to the floor, her eyes wide open, her tongue lolling out of the side of her mouth. She looked stunned, literally: that was the reason for the name "stun gun."

"The gun is electric," Nader explained to me. "It's silent. An electric spike went into this thing's brain and knocked it out."

Mountainous Moss entered the kill floor, dressed like Nader, wearing a hard hat and a white robe covered with blue and yellow apron shields. He crossed the river of blood, then the two men, each clutching one leg to a hand, lifted the unconscious sheep onto a slaughter stand. The stand looked like a laundry drying rack except that it was made of metal and sat low to the ground, the height of a chair seat. The sheep was laid horizontally on her back in its steely embrace.

Nader pulled out a small, curved knife from the extensive assortment around his waist. "Now I'm going to slash, slash,

slash!" he said, laughing, his eyes twinkling. With a rapid thrust of his hand, he slit the sheep's throat.

This was the moment I'd been desperately dreading, and it happened so suddenly. The sheep's neck had been one, and now it was two—ripped in the middle, the windpipe severed, the head now perpendicular to the body, connected only by the strip that was the spinal cord. From the inside, the sheep's throat looked strikingly like a human throat, containing the same patchwork of life-giving pipes. Blood gushed out from the neck in great red torrents, like water from a broken faucet. It splattered my white robe and joined the river on the floor, but remained distinct from it—crimson rather than scarlet, still fresh and full of oxygen.

The sheep's eyes, still eerily wide open, were unblinking, but her legs kicked reflexively. After what seemed an eternity—though it could not have been more than a couple of minutes, the longest of my life—the kicking sputtered to a stop. The legs became limp and folded, and came to rest on the belly like a sleeping kitten's. If not for the open eyes and the bloody throat, the sheep would have looked asleep.

The line dividing life and death is no wider than the blade of a knife. Thin as a thread, it can end in a moment. My hands clenched in my pockets, I could not stop gasping.

Nader proceeded to kill a second sheep in the same manner as the first. Then he hung them up by their legs on hooks. Making an intense, ugly face, he skinned both sheep, trying not to slip in the blood as he slit and tugged off their woolly skin. When he tossed their skin in a corner of the kill floor, it looked disconcertingly like a living animal with its filling missing.

"Skinning is an art," Nader told me, appearing to enjoy that someone was observing him at his most artistic. After the art of skinning came the art of gutting. Nader slashed open the first sheep's chest vertically and thrust in his hands. Because the sheep's body was warm, and the surrounding air freezing, vapor emanated from inside the carcass, forming a white fog—as if the ghost of the animal were escaping from its dismembered body. Nader kept his hands inside the carcass for a minute. "It's so cold here," he said, "but this thing's nice and hot. Want to put your hands inside with me?"

I declined. Nader nonetheless continued to ask me the same question repeatedly over the course of the day.

The kill floor was deafeningly loud and I had no choice but to lean close to Nader to hear him every time he spoke. All the machines and equipment were churning—the scalding tank, the overhead hooks, the hair-removing contraptions—and it felt like we were standing inside an explosive engine.

When Nader extracted his hands from the sheep's insides, he removed with them the liver, spleen, and intestines. They were slimy and disgusting. He slit open the intestines, and a gooey green mulch—previously hay—collapsed to the floor, splashing in the blood. "All these things do is eat," he said. Another slit elsewhere in the intestines and out rolled round black marbles: balls of manure. Next, Nader slit the bladder, and urine poured out in a sudden, stinking stream, just inches from me. I jumped back.

"I'm doing you a favor," he smiled. "Piss is a grease-cutter. It'll make the blood less slippery."

Nader hurled the sheep's internal organs into five large, black rendering bins that lined one wall. At the end of the day, he would drag the bins out to the back of the slaughter plant. At the end of the week, a rendering plant would collect them and crush and grind the body parts to feed back to farm animals.

At Blackwater, I realized for the first time how little of an animal is actually eaten by people. Some body parts are indigestible, like horns and hooves. The animal's outsides—the skin and wool—and the insides—intestines, stomach, liver, lungs, spleen, bladder—are also disposed of entirely. Blood is drained, too. Only the central layer of the animal remains, and from that, bones are removed during processing or at the dinner plate. In sum, only about half of a cow, sheep, or goat is actually eaten by people.

Moreover, there is the feed conversion ratio to contemplate. For every one pound of weight that a sheep, goat, or cow gains, she has eaten several pounds of corn or hay. The land used to grow the corn and hay would be put to exponentially better use—economically, environmentally, and ethically—if it were growing vegetables for people to eat directly. The problem of world hunger can largely be solved by agreeing to eat lower on the food chain, a concept publicized in the seventies in Frances Moore Lappé's *Diet for a Small Planet*.

Standing in blood, staring at death, it is impossible not to view the raising and butchering of animals for food as wasteful and hopelessly inefficient. And yet the numbers of animals killed at slaughter plants in the United States and Canada defy all reason. Eight and a half billion chickens, 239 million turkeys, 112 million pigs, 32 million cattle, and 2 million sheep and lamb were killed for human consumption in the United States in 2013. The slaughter sum total for 2013, over nine billion animals, is a four-time increase from a half-century earlier.

Nader worked in one corner of the kill floor, and in another corner worked another man. He was tall, white, and handsome, with brown hair, broad shoulders, and an athletic frame. He was dressed like Nader and Moss, in white, blue, and yellow—all splattered with red. He acknowledged my presence only with suspicious stares. His name was Victor, and he was a cruel man.

In front of Victor hung two dead sheep. Behind him, in his looming shadow, cowered nine living sheep. Victor had brought all eleven creatures out together from the holding room onto the kill floor, in order to save himself the trouble of repeated trips. The sheep, however, were frantic and panicked.

The gate between the holding room and kill floor was closed, but they attempted desperately to return to the holding room. Contorting their bodies, bending their backs, they tried to fit through the bars of the gate. They could not. Heaving and pushing on their knees, they tried to slip through from underneath the gate. They could not.

Their bodies stood poised to run, but there was nowhere for them to run. Every section of the kill floor was swamped with blood, and they were terrified of it, turning their backs resolutely away from the butchery. Conceding defeat, they crawled under one another, each sheep pushing and shoving to be at the very bottom of the heap, in order to be most concealed—last to die. Trembling, shaking, panting, they huddled together, forming a round woolen blanket.

"If you knew you were going to die," Nader asked me, noticing the direction of my agonized gaze, "wouldn't you be scared?"

"So you think they know they're going to die?"

"Oh, yeah. These things know. They're seeing all the others dying. They get it. They're not as stupid as we think."

The sheep in the holding room were calm; the sheep on the kill floor were frenzied. This was why Nader brought out one sheep or goat at a time from the holding room, and then killed her immediately instead of forcing her to watch the butchery and await her own slaughter, the wait its own manner of torture.

Victor, in contrast, did not mind instilling the fear of death in animals. He did not mind punishing them before he killed them—just as he punished them *while* he killed them. His killing was sadistic and savage.

Instead of stunning one animal at a time, he would stun *two*, one directly after the other, so that he would have to pick up the stun gun once instead of twice, thus saving himself a few seconds of hand movement. He viewed the stun gun as a tool developed not to make animal death easier, but to make his own life easier.

After stunning two sheep, Victor would use his knife to perforate a hole in the flesh of a hind leg of one sheep. He would pass a hook through the hole. Then he would press and hold a button on a remote control, and the sheep on the hook would lift slowly off the floor, her eyes wide open in her stunned state, her tongue lolling out. Dangling by her leg, her body loose and limp, her head pointing downward, her hooves skyward, she would be shifted across a swath of the kill floor to hang above a tub of blood.

There, Victor would "stick" her: He would stick his knife into her chest and wrench it upward into the neck vertically. ("I slit, Victor sticks," Nader described to me. "Slitting is horizontal, sticking is vertical.") Blood would gush out into the tub, splattering also on the living sheep cowering behind Victor, causing them to tremble with renewed terror.

As the first sheep bled, Victor would turn to the second sheep he'd stunned. He would repeat the process exactly, but it would feel very different from the perspective of the sheep. This time, as he would puncture a hole in the thigh, the sheep—though her face would be still—would be kicking her leg. She could *feel*, to a slight or significant extent, the flesh of her leg being gored. Scowling at the inconvenience of her leg movement, Victor would grasp the leg firmly in his hand and continue to pierce it through, as bored by her pain as if he were a carpenter drilling a hole into a cabinet.

The reason the second sheep would kick her leg and the first wouldn't was that, by the time Victor would get to her after finishing with the first sheep, the effects of stunning were wearing off. Stunning induces complete unconsciousness in sheep for only fifteen seconds. After that brief period, they slowly start to regain some measure of consciousness and sentience, the feeling akin to awakening from a head injury.

For the first sheep, the entire process of boring a hole in the leg, passing a hook through the leg, and getting her across the kill floor would take two to three minutes, meaning that the sheep was quite possibly feeling much of what was happening to her. For the second sheep, the time between stunning and sticking was *several* minutes—and in this case, the fact that the sheep was feeling it was evident. She was kicking her leg from the very beginning. But none of this mattered to Victor. Through the leg, he would perforate a hole—and then through the leg, he would dangle the sheep—and then through the leg, the sheep would be shifted torturously across the kill floor before being stuck by Victor.

This excruciating animal agony was inflicted only so that Victor could save himself some hand movement by picking up the stun gun once instead of twice.

After both sheep had bled out, their bodies, which could now be termed carcasses, underwent a long process. First, to loosen their wool, they were dipped headfirst into a "scalding tank," a boiling cauldron whose steam rose into the air like smoke from a chimney. Then, the hook on which they hung transferred them to a large, V-shaped metal contraption that shook them violently to detach their wool. Because their eyes were open even in death, because the scalding tank had wiped off their blood, and because they were shaking wildly—they looked vibrantly alive.

"You regret coming here, don't you?" Nader asked me.

I turned to look at him. Behind his blood-specked glasses, his eyes were scrutinizing me.

The word "regret" was an understatement of the bone-chilling devastation I felt. The day was, without parallel, the worst of my life. My white robe was stained with hundreds of specks of blood, and my nausea was climbing toward a breaking point. My only source of relief was that I hadn't fainted—*yet.*

"I'm just hungry," I lied, forcing out a smile.

"We'll eat lunch soon," he said, smiling back.

Now I had a new reason for stress: hunger was the *worst* possible reason I could have offered for my state. If I'd told Nader I *wasn't* hungry, I could potentially have gotten out of eating the slaughter-plant lunch without arousing suspicion. But now that I'd said I *was* hungry, I'd *have* to eat it. But how could I eat meat—after all this?

Nader and I left the kill floor for a small office, where we would have lunch.

"The slaughter-plant owner is Pakistani like you," Nader told me. "His name is Abdul. His son had a heart transplant. He's one of the only kids in the country to get a heart transplant. . . ."

Abdul? Son with a heart transplant? Who'd told me about heart transplants? . . . Roger!

Roger Harley had told me about his Pakistani friend Abdul with a heart-transplant son—this had to be the very same Abdul, seeing how uncommon heart transplants are in children. The knowledge agitated me because I'd told Roger the truth—I care about animal welfare—and I'd lied to Nader—I want to work at Blackwater. What if Roger had mentioned me to Abdul, like he'd mentioned Abdul to me? And what if Abdul dropped by today? And what if Abdul pieced things together and concluded that I did not, in fact, want to work in slaughter?

I was such a fool. I'd been so naïve. I'd been distributing different stories like Halloween candy—I want to work in veal, I want to work in slaughter—but it was necessary to have the *same* story. The *one* truth, or the *one* lie.

"Does Abdul drop by here often?" I asked Nader, trying to sound nonchalant.

"Yeah. His wife makes lunch sometimes and he brings it for us. This is his office we're sitting in."

Crap. A minute later, the door to the slaughter plant opened, and I held my breath. My back was to the door, and so I could not see, but I could hear. Footsteps approached, and two plastic bags of food were placed on the table in front of me. I looked up hesitantly. It was Moss. Not Abdul.

But my relief was temporary. The bags of food reminded me that it was time for a new episode in my slaughter-plant persona: I would have to eat meat. My nausea magnified.

Moss left to smoke. Victor entered.

Nader opened the bags, and a mouth-watering aroma of fresh, hot curries filled the air. Nader took boxes of food out of the bag and opened them one by one. The first box held bread. The second, rice. The third, lentil curry. The fourth . . . squash curry. The entire meal was vegetarian. Here we were, at a slaughter plant—eating a *vegetarian* lunch.

I was speechless for a second. "Why is everything vegetarian?" I then whined to Nader. I intended by my complaining tone to emphasize that I was a hearty meat-eater, and to thus lend credence to my slaughter plant persona.

"We always eat vegetarian when we eat here," Nader replied. "Guess what I had for dinner yesterday!"

"What?"

"Guess."

"I don't know . . . lamb?"

"No."

"Okay. Beef?"

"No."

"What? Chicken?"

"No."

"Fish?"

"No. I had a big bowl of *cereal*. . . . Most of the time, I eat vegetarian food." Nader paused, then continued in a somber tone: "People ate meat before we were born, and they'll eat meat after we die. But it's not easy to get meat to the table. You know that now."

I remembered something else: Roger had told me that Abdul's brother, who also owned a slaughter plant, was vegetarian. This was revealing: slaughter-plant workers, at least some of them, do not want to eat meat.

Meat is packaged in plastic, sliced into pieces, and then cooked, marinated, grilled, and flavored, until the browned slabs on the plate bear no resemblance to a living animal. But on a kill floor, *all* distance, physical and psychological, is permanently erased, the mind's slate of excuses and rationalizations wiped clean. On a kill floor, one sees precisely *how* an animal is converted from a creature to a carcass, and one detects in the recipe of meat the ingredients of terror and torture.

Most people would not eat animals if they had to wield the knife themselves. Consumers construct and retain a great distance between animals and meat, but slaughter workers cannot. "How deeply seated in the human heart is the injunction not to take life!" Leo Tolstoy, a vegetarian, noted long ago. "So strong is humanity's aversion to all killing . . . an act which is contrary to moral feeling."

Forty-one-year-old Victor was, by his dominating appearance, the sort of man who could have made a lot out of his life, or a little. From his perspective, he'd made a little; from Nader's perspective, he'd made a lot.

Born to Italian immigrants, Victor had a high school degree, a powerful build, and a loving family. He had been a professional football player in his twenties, until knee injuries and surgeries forced him to stop. He got married in his early thirties to a tall blond woman, and he was now father to four little girls who ranged in age from three to nine. He showed me their photos on his phone, warming and smiling for the first time. "That's them on the couch! . . . That's them playing! . . . That's my second daughter. . . . That's my first one! . . . That's me and my wife at a hockey game. . . ."

I found it strange to imagine such a callous man surrounded daily by a gang of giggling girls. "Have your wife or daughters ever seen slaughter?" I asked him. I didn't know what prompted me to ask the question—perhaps it was the fact that his personal life seemed so artificial, so disjointed from his professional life—but my words broke Victor's momentary spell of friendliness.

"No, they haven't seen slaughter," he replied coldly. Tucking his phone into his pocket, he set his fork down and folded his arms on the table. He fixed me with a scary stare, leading me to think again of *The Sopranos*. "Are you a cop?" he demanded. "Are you an undercover cop? Are you writing a report?"

Nader's head snapped up like a vulture's. His eyes widened behind his blood-specked glasses and he looked at me with renewed understanding. "Are you a cop?! Are you a *COP*?!"

I laughed in an effort to defuse the tension. It didn't defuse. Neither man looked remotely amused. "Of course I'm not a cop! Do I *look* like a cop?!"

Without a word, Victor left to return to the kill floor. Nader shifted to occupy Victor's chair. "Talk to me straight," he ordered me with narrowed eyes. "Are you an undercover cop?"

"No!"

"But I'm *sure* you're an undercover cop. You know why I'm sure? Because why else would a beautiful woman come out here?" Thumping his chest, referring to his heart, he announced: "It's all about what we have in *here*. If you're writing a report about all the horror stories you're seeing, I hope you keep my good intentions in mind."

In 2008, Americans were jarred to see undercover video footage recorded by the Humane Society of the United States (HSUS) at a California slaughter plant called Westland/Hallmark Meat Packing. Workers could be seen torturing dairy cows who were too ill or injured to move—called "downer" cows. Workers kicked cows, rammed them with forklifts, jabbed them in the eyes, and applied electric shocks to them.

In response to the public outcry, the U.S. Department of Agriculture shut down the slaughter plant and announced the largest meat recall ever, of more than 140 million pounds of beef, over one third of which had been purchased with taxpayer money and distributed to school lunch and other federal programs. Westland/Hallmark was the second largest supplier of ground beef to the National School Lunch Program, and had even been named a USDA "Supplier of the Year" in 2005.

Only a year and a half after the California Westland/Hallmark debacle, another slaughter plant entered the national spotlight. Called Bushway Packing, this was a bob veal slaughter plant in Vermont. Victor at Blackwater stunned two animals at a time, but workers at Bushway were stunning ten or more, making Victor look like a saint in comparison. Many veal calves were conscious at the time of slaughter. The Bushway slaughter case was different from most because, first, the slaughterhouse was certified as an organic processor, and, second, this time it was not an undercover investigator but a government inspector who blew the whistle.

Public Health veterinarian Dr. Dean Wyatt tipped off the HSUS, which later conducted an independent undercover investigation. Dr. Wyatt was employed with the Food Safety and Inspection Service (FSIS), the agency of the USDA responsible for enforcing slaughter laws. Instead of being praised by his industry for blowing the whistle, Dr. Wyatt, who died a year later from brain cancer, was vilified by it. U.S. Secretary of Agriculture Tom Vilsack described his behavior as "inexcusable."

The Bushway case humiliated the USDA because it showed the systemic collusion between government and industry. Footage revealed USDA inspectors ignoring the egregious cruelty before them, and suggesting to workers that they abuse calves only when Dr. Wyatt was not present. Moreover, in the wake of Westland/Hallmark, FSIS was not urging extra vigilance, but was instead, according to Dr. Wyatt, instructing inspectors to report the most serious violations only through faxed documents, which fall outside the computerized database of "noncompliance reports" accessible to the public.

Dr. Wyatt's testimony to Congress served as an indictment of the American slaughter system. "People have asked me why I would risk ruining my career by testifying," Dr. Wyatt stated. "I would respond by quoting Abraham Lincoln who said 'to sin by silence, when one must protest, makes cowards of men.' . . . I truly believe that the USDA inspector is the only advocate animals have in slaughter plants. When we turn our backs on the helpless, when we fail to speak on behalf of the voiceless, when we tolerate animal abuse and suffering, then the moral compass of a just and compassionate society is gone."

Dr. Wyatt detailed multiple humane slaughter violations he'd witnessed at the two slaughter plants where he'd been stationed in his career, including Bushway. Both places "deliberately and willfully ignored" his suspension orders, but FSIS sided irrevocably with them, "chastising" and "berating" Dr. Wyatt, and threatening him with demotion, termination, and transfer. Dr. Wyatt concluded his Congressional testimony by urging FSIS to treat "the consumer as its client, not industry."

The Government Accountability Office (GAO), the audit and evaluation arm of the American Congress, has written multiple reports urging a restructuring of slaughter inspection. Upon analyzing survey responses from more than 200 FSIS slaughter inspectors in 2010, the GAO concluded that surveyed inspectors did not "suspend plant operations or take regulatory actions when they appeared warranted." Moreover, half the inspectors could not even "correctly answer basic facts about signs of [animal] sensibility."

After lunch, I returned to the kill floor for round two. I saw then with glaring clarity how severely I'd endangered my life by coming to Blackwater.

First, there were Nader and Victor, both suspicious of me, both dangerous. Nader was mentally unstable—I was certain of it—and Victor was openly cruel.

Second, Nader, Victor, and Moss all wore yellow, construction-worker hard hats to prevent head injury. A safety sign at Blackwater's entrance warned that a hard hat was a requirement for every individual on the kill floor. I, however, had not been given one. I worried that a heavy, overhead metal hook would fall onto my head and debilitate me permanently, as it had Nader. "A hook fell on my head once," he told me, "and it took me a while to move again. I've been wearing a back brace since then. I can't stand up straight if I don't."

A further peril was propane. The scalding tank used to loosen wool was heated from underneath by a thick propane pipe, and whenever Nader skinned animals, he accidentally stepped on the pipe and pulled it out from underneath the scalding tank. The fire then lay *in* the blood on the floor, lighting it yellow. With this propane flame, and with the barrels of propane lining the rear of the slaughter plant, the place was extraordinarily flammable.

Another source of discomfort, despite the fire, was the cold. In spite of my wearing several layers of clothes and two pairs of socks, as Nader had instructed, my toes were numb. The only way to keep them warm was to continually move them in the blood, but the blood was slippery, and I didn't want to fall.

Little seemed to have changed in slaughter since *The Jungle.* "There was no heat upon the killing beds," wrote Upton Sinclair. "The men might exactly as well have worked out of doors all winter. . . . Now and then, when the bosses were not looking, you would see them plunging their feet and ankles into the steaming hot carcass of the steer. . . ."

Nader didn't insert his feet into carcasses, but he did like warming his hands in them.

Finally, the pants of my men's-size scrubs fit me no better than a blanket—and fell off. It was a testament to my overwhelmed, light-headed state that I didn't even notice. I learned about my wardrobe malfunction only when Moss broke into a peal of laughter, and Nader joined him. I couldn't care less, for I was wearing pants and thermals underneath, and my attire was the least important issue of the day.

Nader helped me re-tie my scrubs and robe. The robe literally went around me twice. "You have blood on your face," he informed me casually, his face only inches from my own.

Horrified, I began scrubbing my face frantically against my coat sleeves. Then I noticed that my sleeves were already drenched in blood. I stilled into a repulsed stupor.

"Don't worry!" Nader reassured me, noticing my agony and assuming it stemmed from vanity. "You look good with blood on your face. I have blood all over my face too."

I stood stoically next to Nader as, one by one, he dragged sheep and goats out from the holding room. He stunned them and slit their throats. Each sheep and goat died a different death. Some died slowly, some rapidly, some throbbed mildly, others shook wildly.

As Nader hung and gutted the animals, I noticed that at least a third of them had internal organs covered with oozing, pus-filled abscesses. The liver tended to be most affected, looking like a flat, dark expanse pulsing with volcanoes. The spleen and intestines were also affected, smeared with creamy, crumbling white and yellow pus. Even these pus-oozing innards, however, were hurled into rendering bins, to be eaten by farm animals. In cattle, abscesses are the outcome of excessive corn in the diet, as the bovine rumen has not evolved to digest corn. In sheep and goats, the reason for abscesses hasn't been studied as much but is likely to be similar.

It was beyond the capacity of my mind to imagine the effects diseased body parts would have on the health of the animals that would eat them, and it was more difficult still to imagine their effects on the health of the people who would then eat those farm animals. It seemed doubtless, however, that the convoluted, disease-filled food chain would have effects. People would not be consuming the abscesses themselves, but they would still be eating the animals from which they came, and also the animals that ate the abscesses.

Taking seriously my stated aim of starting a slaughter plant, Nader started treating me like an apprentice. "Catch a sheep," he ordered me in the holding room. I demurred. He insisted. I attempted and failed. He then commanded me to stun a sheep. I fired the gun, but missed on purpose, shooting at the floor instead of the sheep. Nader snatched the gun out of my hand and stunned her.

Nader and Victor, I noticed, had no uniform method of stunning, and seemed to, in fact, be attempting to introduce diversity to the practice. They would place the gun directly on the skull, or they would shoot from a few inches away, or they would shoot the front of the skull, or the back, or the side, whichever happened to be closest.

"Stunning's often not done properly," pastoral farmer Roger Harley had told me, speaking from his experience auditing slaughter plants. "Animals should be stunned at a precise angle at the top of the head. But workers often don't have a clue, and, even if they do, they don't care. They often stun the wrong part of the head. I've seen cattle standing there after being stunned. And then they're stunned again—but they're still standing there. That means the stunning isn't even close to being in the right part of the brain. Also, after animals are stunned, they should be killed pretty quickly, because otherwise they start regaining consciousness. But killing's often *not* done quickly. In my audits, lots of animals are partially or fully conscious when they're slaughtered."

Victor looked up at me suddenly. I looked away. He started walking toward me.

What could he want? Did he want me to leave? Could he see something in my eyes?

"Can you do me a favor?" he asked. "Can you go upstairs and get Maara?"

Maara was a short, fat, black woman from Trinidad with a slow walk and a lilting accent. She was the government inspector assigned to Blackwater.

Government inspectors are assigned to slaughter plants in the United States and Canada to ensure that animals are not mistreated, diseased, or unsafe for human consumption. Each of these functions is distinct, requiring its own skill set and knowledge, but the same person is required to perform all three of them. Inspectors spend almost all of their time on just one of the three: meat safety.

"The inspector [at Bushway] spent ninety-nine percent of his time doing carcass inspections," noted Dr. Wyatt in his Congressional testimony, "and was unable to do his humane slaughter and other duties properly. The plant management loved this."

Though Maara was, by law, required to be present on Blackwater's kill floor from the moment slaughter began until the moment it ended, she was hardly ever there. She spent almost all of her time comfortably ensconced in the office upstairs, checking her e-mail. She deigned to descend to the kill floor no more than once every two hours or so, when she quickly stamped carcasses and pushed them into the "chill room." During those limited intervals when she was actually on the kill floor, 99 percent of Maara's time was spent on carcass inspections.

Victor had asked me to fetch Maara because more than two hours had passed, and she had not yet appeared to stamp carcasses. This posed a food safety hazard, because carcasses had to be chilled, and they could not be pushed into the chill room without her round, blue, waterproof stamps. Obeying Victor, I requested Maara to descend to the kill floor. She arrived close to an hour after my request, leading Victor to sarcastically mutter, "You finally came, huh?"

Maara's carcass inspection consisted of examining the lymph nodes on the sides of the head, then slicing two slivers of flesh off the face, her knife eerily grazing the eyes, which were still wide open. "I'm doing this to see if they have abscesses," she drawled to me. It did not matter whether there had been abscesses on the animals' innards; so long as they were not on the face—and they never were—Maara approved the carcasses for human consumption.

She was the epitome of the sort of inspector described in *The Jungle*. Carcasses "had to pass a government inspector, who sat in the doorway and felt of the glands in the neck. . . . This government inspector did not have the manner of a man who was worked to death. . . . If you were a sociable person, he was quite willing to enter into conversation with you."

Maara—not "worked to death" by any means—was quite willing to enter into conversation with me. "I became a slaughter inspector because it pays very well," she told me in the tone of imparting career advice. "I make thirteen hundred dollars a week."

Although Maara was a government employee, she, like other slaughter inspectors in Canada, was actually paid *by* the slaughter plant. This meant that Maara—earning far more than hourly workers like Nader, Victor, and Moss—had more incentive than any other person except the owner to keep Blackwater breathing and living.

"If an inspector shuts down a plant," Roger Harley had explained to me, "they're out of a job, because they were being paid by the slaughter plant. For an inspector to shut down a plant is like shutting themselves down, putting themselves out of business. So, what always happens is, the inspector becomes one of the boys and turns a blind eye to everything."

"Have you ever shut down a slaughter plant?" I asked Maara.

Startled at the question, she dropped the lymph nodes that she was perfunctorily examining. "The places I inspect are all good," she said. "Like this place. This place is good. As an inspector, you want slaughterhouses to like you, and you want to like them, too. That makes life easier for everyone."

Except, of course, the animals.

I asked Maara questions about stunning. "All I know about stunning," she drawled, "is that after you stun, you have to kill the sheep in fifteen seconds. By twenty seconds, it starts coming back."

The irony of the day was that this iron law of stunning was being violated just as she was describing it. As always, Victor had stunned two sheep at a time, but it happened to be at that very moment that he was boring a hole into the leg of the second sheep—not seconds, but minutes after stunning. The leg was kicking violently. Maara did not see this, however, because, as always when she inspected carcasses, she chose to turn her back to the killing.

She spent her time at Blackwater resolutely avoiding the kill floor, and she spent her time on the kill floor resolutely avoiding the killing.

"I'm going to the holding room to smoke and take a break from all the killing," Nader said.

In the holding room, Nader told me more about himself, and he also showed himself to me, starting with his hands. On his left hand, he wore a green glove encased in a steel grid to protect his hand from his own knives. He took his glove off and stretched his hand with a wince. "I have carpal tunnel in my left hand," he said. Over his right wrist, Nader wore a wide steel bracelet to stabilize a joint that had, over half a lifetime of knife-wielding and animal-lifting, become loose and lax. Both his hands—wrists to fingers, palm and back—were crisscrossed with a quilt of scars.

Most industries have become safer over time, but slaughter has become more dangerous. Injury and illness rates are among the highest of all industries. Men work in slaughter when they can find no other work, but it also makes them unfit for any other work. Fifteen to 30 percent of slaughter workers suffer cuts, strains, traumas, falls, fractures, musculoskeletal disorders, back injuries, hernias, facial disfigurements, burns, carpal tunnel, severed fingers, severed arms, amputations, and other injuries.

Today, half of American slaughter plants employ fewer than ten workers, like Blackwater. Another quarter employ between ten and forty workers. Hundreds of American slaughter plants, however, employ more than five hundred workers, and some employ more than a thousand. These immense killing factories supply the vast majority of American meat.

Close to two fifths of the 80,000 slaughter workers and meat packers in the United States today are Mexican, and an undetermined but considerable proportion of them are illegal aliens. A startling four fifths of slaughter workers have not completed high school. Their median pay is eleven dollars per hour, a paltry sum when set against the daily dangers of their job.

"[Slaughter] workers are exposed to noise, dangerous equipment, and hot or cold temperatures," states the U.S. Bureau of Labor Statistics. "Lifting heavy pieces of meat and doing repetitive cuts all day make the work physically demanding. . . . In addition, processing plant floors are often slippery, causing workers to fall. . . . Working conditions in most processing facilities are physically demanding and often very difficult and unpleasant, resulting in high job turnover."

Nader was extraordinary for having lasted sixteen years in the industry. Most slaughter workers—underpaid, overworked, injured—last less than a year. One third quit in their first month on the job.

Nader's entire body was a throbbing mass of aches and pains, but his mind was even more bruised than his body. "I'm on lots of meds," he said, pointing to his head. "If I don't take my meds, things get messy. I'm not allowed to drive because of all the meds I take. I have paralyzing schizophrenia and I'm on anti-psychosis meds prescribed by a psychiatrist. I have to see a psychiatrist once a month. I also have to see a probation officer once a month."

I didn't have the courage to ask Nader what he was on probation for.

"My anti-psychosis meds make me sleepy every night. They knock me out at seven—I call them my 'knock-out' meds. I have other meds, too. I have meds for mood swings. I have meds for ADHD. And I used to be a crack addict. I'm not anymore, but when I dropped out of school at sixteen, I joined gangs and became a crack addict . . . I'm on disability [insurance] because of my mental problems."

I didn't know what to say. I'd never met anyone with so many psychological issues.

"The disability [insurance] helps me pay for my food," Nader continued, "but it's not enough. It's just my bread and butter. So I keep working, even though I'm not supposed to work. If anyone reports me working here, it won't be good. I'll lose my job . . . I think my problems are because of all the killing. My job is very violent. I'm yelling at animals all day. I'm pushing them around. I'm killing them. It doesn't feel good. But I can't change my job; I don't know any other kind of work . . . I had a lot of nightmares when I first started killing."

Nader's experiences were, tragically, standard for his industry. Several studies demonstrate that slaughter workers incur mental trauma and resort to drug abuse and criminal acts. The mental disintegration of workers also leads to community disintegration. A 2009 study found that "slaughterhouse employment has significant effects on arrests for rape and arrests for sex offenses," in addition to arrests for violent crimes in the area. A 2004 book called *Slaughterhouse Blues* found that a Kansas county suffered a three-fold increase in child abuse and a 130 percent increase in violent crimes within five years after two slaughterhouses opened. Other counties with large slaughter plants show similarly stark outcomes.

Nader knew no work other than killing, but his killing of animals was killing him slowly, bleeding him drop by drop, from the inside out, as if his knife were pointed inward. He walked a daily tightrope between life and death, between sanity and insanity. I felt sorry for him and angry on his behalf. War veterans receive benefits; shouldn't slaughter workers, too? Aren't they at war? Isn't a kill floor a daily battle, a perpetual blood bath?

"You owe me," Nader said. "You really owe me. I let you come here. I could get in trouble for that."

I pulled out all the cash I had, a grand sum of only thirty dollars, wishing I'd brought more. The look on Nader's face was not appreciation but affront. He looked away from the money like it was bribery.

"I like helping people," he said grandly. "I don't mean you owe me *money*. You owe me *more* than money. I'm a loser, and I don't have any friends. I don't have a girlfriend. You owe me a *date*. You're like a dream come true. I get so lonely. I want to settle down with someone and get married and be happy."

"Oh."

"And just so you know, I come from a good family. My cousin's wife's father had a big job at Chubby's a few years ago. It's hard for me to find a wife as a butcher. No one wants to marry a butcher. But I can tell you're different. You're fine on the kill floor. You have a strong stomach. Other women scream, *'Aaah! Aaah!'* You don't. I can help you some more. I can lighten you up. I know some big shots in the game."

"What do you mean you can 'lighten me up'? What big shots? What game?"

"I mean I can set you up with somebody."

"Like who?"

"I'm set with some big dealers."

"What do you mean?"

"I mean I can help you get work in *slaughter*," he spelled out impatiently. "I can get your foot in the door. I can get you in, set you up. And we can also raise animals together on the side. You can do the lambing, then I'll do the killing, then you can do the selling to the restaurants. I've always wanted to have a family business. We can wheel and deal *together*."

"Come on, ladies!" Nader said.

"Ladies" referred not to me or Maara, but to the four sheep and two goats left in the holding room at the end of the day. He dragged them all out together by their hind legs, impatient to end his day. "I'm going to do these last animals halal," he informed me. "They're for a halal restaurant."

"Halal" is the term for Muslim ritual slaughter, as "kosher" is the term for Jewish ritual slaughter.

"I'm Muslim, but I don't like halal," Nader continued. "The only thing halal means is that you aren't allowed to stun animals. But studies show stunning is better. Back in the day, people used to kill animals by hitting them on the head with a bat. We don't do that anymore. Stunning is good, it's a part of technology, but Muslims still don't do it. How would *you* feel if you were hit on the head with a bat?"

Everywhere the four sheep and two goats looked, everywhere they stepped, everywhere they smelled, they were surrounded by raw blood and rotting innards. It made them frantic.

Their deaths were a world apart from those of the stunned animals earlier. As they were laid horizontally on the slaughter stands, they kicked their legs wildly and defecated copiously. When Nader slit their throats, after saying the perfunctory halal words, "*Bismillah, Allah-u-akbar*"—"In the name of God, God is great"—they blinked, blinked, and blinked. Blinking is a definitive indication of consciousness.

Their legs continued to kick and their severed windpipes fought to draw breath. Their heads were now attached to their necks only by the spinal cord, but they still *lifted* their heads. Their mouths moved and a *hiss* emerged from between the lips—a sharp, snake-like sound.

"Wanna know what these things are saying?" Nader asked me. He imitated the inimitable sound, like he was being strangled, and hissed, "Why are you killing *meee*?" He laughed.

Nader then did something I could never have imagined. Announcing, "I don't have time to wait for these things to die," he sliced off their forelegs at the knee—*while* they were still conscious: gasping, hissing, blinking, kicking, *looking* at him. Throbbing and convulsing, they kicked so much that they fell off the slaughter stands, headfirst into the pool of blood, some of it theirs. They continued to thrash in the blood, their back legs moving like they were running for their lives. Eventually, the legs ceased, the head lowered, and they breathed their last breath.

The American Humane Methods of Slaughter Act and the Canadian Meat Inspection Regulations exempt Jewish and Muslim ritual slaughter from the requirement of stunning in order to protect "freedom of religion." Many European countries, in contrast—including Sweden, Norway, Switzerland, Iceland, and Denmark—require stunning for *all* farm animals, regardless of religion. In 2014, defending his government's decision to mandate stunning for all animals, the Denmark Minister for Agriculture and Food stated on television: "Animal welfare comes before religion."

One goat at Blackwater died differently from all the others. As Nader and Moss lifted her onto the stand on her back, she kicked Nader. Her hind leg rose like a stick and struck his upper lip with such force that his hard hat flew off and fell a distance away. (Worker protection, in addition to animal welfare, is an important reason to stun farm animals.)

"FUCKING THING!!!" Nader screamed. He slashed the goat's throat savagely, not bothering with his perfunctory one-line prayer this time. He sliced off her forelegs with relish. She hissed, she thrashed, she throbbed, she collapsed into the blood on the floor.

I couldn't take it anymore. Shaking, I walked off the kill floor.

"Do you think she's had enough?" I heard Victor ask Nader as the door closed behind me.

I didn't hear Nader's answer, but the right answer would have been yes.

An hour later, by the end of the day, the kill floor was cleaned, scrubbed, sanitized. There were no traces of blood, no wails of animals, no screeches of machinery. The place looked unrecognizable, as if nothing had happened there.

But something *had* happened. A conversion had occurred. In the morning, all animals had been in the holding room, alive. In the evening, all animals were on hooks in the chill room, dead.

Nader took me to the building next door, which was a chicken slaughter plant, also owned by Abdul. He could sense my fear in being alone with him, just as well as he could sense fear in sheep. "Don't be scared of me," he told me repeatedly. "I'm not going to hurt you." His assurances only agitated me further.

The ice along the way was slippery. "I'll catch you if you fall," Nader offered. He extended me his arm. The gesture was comical because it was more suited to a ballroom than a slaughter plant. I refused his arm at first but then consented to it when I came close to falling. "You look so fragile," he said, "like if you fall down, you'll break all your bones."

Fragile. That word again. Just the day before, I'd received an e-mail from a friend to whom I'd vaguely mentioned that I was these days visiting farms. "You look so fragile, dear," she'd written, "that I hope you're not doing more than picking tomatoes." The word "fragile," I realized as I walked on Nader's arm, had once been an accurate description of me, but was now like an old, shrugged-off skin. I was no more fragile than a cactus.

The chicken slaughter plant was not yet open, but its construction was complete. Large, sleek, steel vats and overhead hooks filled the place, poised to slay 20,000 chickens a day.

"You did well today," Nader told me with a smile. "We're going to be opening up this chicken place soon, and I'm going to be hiring. Do you want to work here? With me?"

"I'll think about it."

Nader's face fell at my lack of enthusiasm.

Moss and Nader were my ride to the train station (which was why I'd stuck around as long as I had to begin with). Moss drove the blue van, Nader sat in the passenger seat, and I sat in the back. Because of his "brain meds," Nader wasn't permitted to drive. Nader sang along to the radio with as much mirth as if we'd spent the day at a park. When Moss eventually stopped the van, it was not at the train station, but a restaurant.

"We're making a quick delivery before dropping you," Nader told me. He collected a dozen freezer bags of meat from the trunk of the van and took them into the restaurant. It was a small, Jamaican takeout place with bright colors and no customers. A sign next to the door read: PEACE JOY LOVE.

I spent the next days holed up in my apartment, crying, reeling, recovering. I'd never been this shell-shocked before. It didn't help that Nader called me obsessively.

"Did you have nightmares? I've been killing so long that I don't, but you probably did. I told you you would. . . ."

"I hope you're not mad at me. Are you? Maybe the day was too much for you? Anyway, I'm going to take my knock-out meds at seven, so call me before that. Call me, okay? *Okay*?! Or maybe I'll call you again. . . ."

"The government inspection's done at Blackwater, and we're opening up the chicken place in a few weeks. Want to work here with me? . . ."

"There's a wedding coming up. Want to go with me? . . ."

"Have you forgotten about your day here? Don't *ever* forget about the meat industry. . . ."

"Remember what I said? I can connect you to some big shots in the game. . . ."

"I made a lot of money killing last week and I want to take you out for dinner and a movie. Call me back? *Please*?"

"Somebody reported me working here. Now I can't work here anymore, 'cause I'm on disability, and I'm not supposed to be working. I don't know what to do. . . ."

"I moved because I was reported. Now I'm working at a cattle slaughter place. Maybe I can make arrangements and call you later. I can get you a tour here, then we can go to my place after. Yeah? Okay?"

"Call me back, okay? *Please*. . . ."

PART II

CHAPTER NINE

THE WONDROUS ISLANDS OF INDONESIA

COCK-FIGHTING AND STRAWBERRY-PICKING

At Blackwater slaughter plant, I'd reached some sort of internal limit. When I'd wavered at the door to the kill floor, I'd known that something in me would be altered permanently. It was.

Wall Street now seemed like a distant, faraway mirage, forever out of reach. I also felt alienated from friends and family. I felt that their lives were lived, both intentionally and unintentionally, behind a curtain of pretense. I felt that society itself lives behind a giant subterfuge. The facts of industrial agriculture are so removed from the day-to-day lives of most people that they might as well be occurring in a separate universe.

I'd found lots of troubling issues in Canadian treatment of farm animals, but I'd also detected some solutions in the midst of the mess. I'd learned the value of pastoral farming. I'd learned the importance of proper inspecting, both in farming and slaughter. I'd learned the necessity of good laws. But there was much more for me to learn still. It was time to expand my horizons.

I'd seen everything there was to see in Canada—dairy cows, egg-laying hens, pigs, turkeys, chickens, sheep, beef cattle, veal calves, and slaughter. I'd have to go further to learn more. I decided that I would travel to some faraway place where I knew no one—from where I'd never even *met* anyone. There were not many such places in the world. The options, as I narrowed them down, came to be Antarctica and Indonesia.

I preferred Indonesia. I did not know how Indonesians looked, how they dressed, how they lived, what they ate. What might the similarities and the differences be between agriculture in the east and the west? Could Indonesia point me in new directions, toward new solutions for farm animals? Would I discover positive practices that could be applied globally?

I bought a one-way ticket to Bali with my identical twin sister, Sofia.

Sofia and I are best friends. Both of us attended Dartmouth College, and both of us graduated with economics majors. Both of us worked on Wall Street, at investment banks directly across the street from each other. Both of us lived in New York, then Toronto, sharing an apartment in both cities. Our life trajectories were, in other words, perfectly identical throughout our lives, with one exception: my stint at the Miller dairy farm. That one stay, that one brief divergence in our paths, thrust us, for the first time in our lives, in entirely different directions. Sofia had a successful financial career; I would be traipsing about Asia indefinitely.

We landed in Indonesia on a humid midnight in June. Our first observation was that Indonesians have the same medium-brown skin color as we do. The bed-and-breakfast we'd reserved online had not bothered to remain open for us, and so we had to scramble to find a place in the middle of the night. We eventually did, and fell asleep in exhaustion. When we awoke the next morning, at dawn, it was not to an alarm clock, but to birdsong.

The song was a feast for the ears, a syrup-sweet orchestra of birds with vibrant scarlet and violet plumage, who sang sometimes in solos, at other times in duets, and often in full ensembles. We'd never heard such birdsong before—we hadn't even known such stirring music existed in nature. It became, to us, an audible drug throughout our stay. The more we heard, the more we pined for.

Indonesia is described as an archipelago, a cluster of thousands of islands. Sofia and I stayed in villages on four islands. Among the first of them was a village called Tetebatu (*Tay-tay-baa-too*), on the Muslim island of Lombok. Though our rustic accommodations left much to be desired, with stains on the beds and staring geckos on the walls, the delicious meals the family grandmother cooked us every day—sizzling, spicy mushrooms, using an extraordinary variety of local mushroom she grew herself—made up for them.

We spent a day walking through the streets, fields, and forests of the village with a tour guide called Badrun (*Budd-run*). Standing at less than five feet in height, he was thirty-two years old and had outward ears, cunning eyes, and a grimace of a smile. Like most men in Indonesia, he told us, he was a chain-smoker. Unaccustomed to tourists of our color and size, he commented, "Other tourist very white and very pat." (In Indonesia, the letter "f" is often pronounced as "p.")

As we walked about, we saw that most Indonesian women wear head-scarves, and men and women both wear sarongs, which are long, skirt-like cotton sheets in plain or plaid patterns. Most people have a cell phone (or two) and most households have electricity, running water, a television, and a motorbike or two. Unlike many other developing countries, Indonesia makes schooling a requirement for children, because of which fact 90 percent of Indonesians are literate.

Located in the Southeast Asian region, Indonesia is a vast nation whose history has been shaped by trade with Hindu kingdoms, Buddhist dynasties, Muslim sultanates, and Dutch colonialists, all of whom arrived at different periods for different purposes, and overlaid one another's influences. Hundreds of indigenous languages continue to be spoken in the country, but there is one national language, called Indonesian, which forms the tongue of school and business.

Badrun, Sofia, and I walked through colorful fields and gardens boasting bananas, bamboo, coconuts, cassava, coffee, cacao, peanuts, strawberries, chili peppers, and bell peppers. Indonesia, because of its moist tropical climate and fertile soil, boasts an abundance of flora and fauna. We also traipsed through verdant rice fields, carved out as lushly beautiful, postcard-perfect steps on hills. Tall, slender rice crops swayed like dancers in the breeze. In the pools of water around their base swam fish and frogs, introduced there by farmers to consume insects, and thus a far more natural and environmental solution than artificial pesticides and chemicals.

Almost every family in Badrun's village had a flock of children and a flock of chickens, the latter numbering anywhere from a half-dozen to two dozen. Roosters were large and extravagant, with a cape of bronze and gold feathers, and a dark, fan-shaped tail that shimmered green and indigo in sunlight. Hens were smaller and simpler, their feathers a plain brown color meant to provide camouflage and protection from predators as they cared for chicks.

Chicks were shades of buttercup, beige, and brown, and they twittered as they trailed their mothers in an enthusiastic queue.

Like Sofia and me, the chickens in Indonesia had a bed-and-breakfast setup. They ate their first meal of the day, rice and corn, at home, and they passed the morning and afternoon foraging in the village for seeds and worms. Using their keen sense of direction, they always returned home before sunset, roosting during the night on low-hanging branches of trees.

Hens were clearly maternal creatures. They would not let us get close to their chicks, instead turning around and running off. Sofia and I asked Badrun how it was that cats in the village walked past chicks with as little interest as if they were pebbles, giving them not so much as a second glance. "Chicken will bite cat," he said. Indeed, on those occasions when we did see a cat get too close to chicks for comfort, the hen fluffed up her feathers and pointed her beak at the cat like a dagger.

An interesting incident occurred. Though most hens had the run of the village—as free as the birds sailing the skies—we passed one hen in an enclosure with six chicks. She was clucking to them gently and they were tweeting back to her. The scene was as charming as any could be, but when we passed the enclosure again, we saw only the hen. Where were the chicks, we asked Badrun.

"I don't know," he replied, rolling his eyes. "I boring already."

He meant he was bored. I squatted in front of the hen. She met my eyes, her glance cold. There was no indication that chicks had ever even existed in the enclosure. I sat there scratching my head, ready to concede defeat and re-evaluate my reality. The chicks had somehow been a figment of my imagination. But then, just as I was about to walk away, a tiny foot emerged from underneath the hen's feathers. A little golden head peeped out. And then a little chocolate head, to the other side. Witnessing my large form, the two chicks vanished again underneath their mother's feathers.

It was wonderful. The chicks knew when to hide, how to hide, how long to hide. Badrun had a flock of a dozen chickens himself, he informed us with a yawn.

"How often do you eat chicken?" I asked him.

"I don't know. It depend. Maybe once in one month. Rich man in Indonesia eat chicken once in one day, poor man eat chicken once in one month. I poor man. I eat my own chicken—local chicken. White chicken, I don't like."

"White chicken," Badrun explained, is a generic term in Indonesia for the variety of white, industrially bred chicken that has come to the country from the developed world. White chicken is the sort of sickly, sedentary chicken I'd seen in Terry's broiler operation in Canada. The mahogany and auburn chickens we were seeing in Badrun's village are called "local" chickens in Indonesia, because they are local to Indonesia and, until just the last decade or two, were the only sort of chicken present in the country.

"Do people in your village eat white or local chicken?" I asked Badrun.

"Local. I did not even hear of white chicken until tourist start to come to Lombok island. My priends in this village don't like taste of white chicken; only tourists like. Also, white chicken—you have to give pood to it three time in one day. It cannot get own pood like local chicken. It is too pat; it cannot move. And it die past, very past. Anything happen, it get scared, it die. In one or two month, it die. Local chicken not die like that. Local chicken very strong."

When it came to both taste and cost, Badrun and most people in his village preferred local chickens to white ones. White chickens had to be fed three times a day; local chickens just once a day, and even that was unnecessary—a morning meal was furnished them mainly in order to lure them home at night in anticipation of breakfast the next morning.

I asked Badrun what Indonesians eat other than chicken.

He counted out five items: rice, vegetables, eggs, soy, and meat.

Rice, he said, is by far the staple of Indonesia. It is eaten for breakfast. It is eaten for lunch. And it is eaten again for dinner. It is eaten by people. It is eaten by birds. It is eaten by dogs, and even by cats. During our village tour, Sofia and I were amazed to come across a cat gobbling up a hearty bowl of rice.

After rice, Badrun said, the next four items occupy equal places on the Indonesian plate. Vegetables are a given. Eggs are considered a bonus—a fried egg is often tossed like a lopsided crown on top of a plate of rice, a dish called *nasi campur.* Tofu is common, as is tempeh—a form of tasty, fermented soy cake that originated in Indonesia hundreds of years ago, the only major soy food to have originated outside of China and Japan. As for the eating of animals, the story is more complicated, as always.

Meat is religious—or, often, sacrilegious. Four fifths of Balinese do not eat cows because, being Hindu, they view them as sacred. More than four fifths of Indonesians outside of the island of Bali, in contrast, are Muslim, and do not eat pigs because they view them as filthy. In other words, cows are too good for Hindus to eat and pigs are too bad for Muslims to eat.

Cows and pigs occupy wholly different standings in Indonesia than they do in the West. The religious politics of eating animals apply to most of the world's population, in fact. Close to one quarter of the world's population is Muslim and foregoes pigs. One seventh of the world's population is Hindu and foregoes cows.

Indonesians, like most other Asians and Africans, do not drink milk. Milk contains a sugar called lactose, which is broken down in the digestive system by an enzyme called lactase. But production of the lactase enzyme ends after an infant's breast-feeding period. This is a condition called lactose-intolerance, whose symptoms range from a rumbling stomach to explosive diarrhea. Two thirds of the world population is lactose-intolerant. *Not* drinking milk, in other words, is biologically more natural than drinking milk.

Badrun, Sofia, and I also passed by some cows.

At five feet in height, the brown cows, called Bali cattle, looked miniature in comparison to their North American counterparts. They lived in groups of two to five in fields—sometimes tied with long ropes to trees—or, more often, in individual wood sheds. Badrun explained that they were tied in order to prevent them from "eating everything in the village."

It was disappointing to see that cows were unable to explore the stunning green fields and valleys all around them. But as it turned out, this was not the full picture. "Those are very big dogs!" Sofia exclaimed.

Toward us from far ahead on the village path bounded three creatures, their forms unclear, their excited, energetic pace suggesting that they were dogs. As they approached, however, we saw that they were not dogs but . . . calves! Far behind them walked a man with three cows on long leashes. The man and the cows passed us, but the calves took an alternate, more adventurous route: they leapt over a ledge of fallen bamboo and disappeared into the brush.

Sofia and I watched them open-mouthed: we had not known calves could run or jump like that, and we would not have believed it had we not seen it.

"Cows with man are mothers," Badrun explained. "Babies do not need rope because they stay near mother. Man is walking cows. People walk cows every day for half-hour or one hour, so cows don't get boring."

Badrun meant so cows don't get bored. I found it enthralling to see that people in Indonesia walk their cows, as Americans walk their dogs. Not only do they keep their cows and calves outdoors year-round, but they also make sure that they get daily stimulation and exercise.

The most enjoyable part of the village walk for Sofia and me was to come across, in a grassy patch, an odd squad of two plump puppies and three yellow ducklings. One of the puppies nipped my finger, then curled up on my lap for a nap. Badrun was staring at the puppy, so I approached him with her.

He jumped back, his lip curling with disgust. "Muslim don't like dogs! I am Muslim. I don't touch dogs. I have dog; I never touch it."

"Then why do you have a dog?"

"For security, in case there is thiep. Dog is not pet. It does not have name. It does not come in house."

Badrun's opinion paralleled that of most Muslims, for dogs are viewed as dirty in Islam (the only dirtier animal is the pig). Traditionally, dogs are considered acceptable to work with—they can protect the house—but unacceptable to fraternize with—they can never enter the house. They are viewed as "the lowest of the low," as described in a book titled *Animals in Islamic Traditions and Muslim Cultures*.

Over the course of our stay in Indonesia, Sofia and I would notice that in some Muslim villages, there are no dogs to be found at all, since residents do not want them there.

Sofia and I proceeded next to a city called Sengiggi (*Sen-gee-gee*), on the same Muslim island of Lombok. There, in the hotel lobby, we met a professor.

His name, Muhajir (*Mu-haa-jirr*), means traveler in Arabic, and it was given him by his parents because they were migrating from the Indonesian island of Java to the island of Sumatra when his mother was pregnant with him. If one were to look at Professor Muhajir, one would guess his profession accurately, for he wore glasses, had a square form, and was filled with an endearing optimism about the world.

Professor Muhajir had obtained a Ph.D. at the University of Kentucky, where he'd resided for five years. "Well, I like America very much," he professed. A soil scientist in Indonesia, he'd traveled to thirty countries for environmental conferences in the previous ten years, and he was at this hotel for a conference as well, one of more than a hundred professors who represented all the major islands, languages, and religions of Indonesia.

Professor Muhajir invited Sofia and me to attend the concluding conference breakfast with him. We happily agreed. The two-hour-long breakfast conversation with Professor Muhajir and the other environmental professors at his table was illuminating—in that it was *not* illuminating.

Since Professor Muhajir was a soil scientist, we asked him about all the plant burning we'd seen during our village tour with Badrun. We'd passed several stacks of smoldering rice stalks, their smoke billowing up to the sky. Why did villagers set fire to their plants after harvesting them? Was all this burning not harmful for the environment?

"Well . . . setting fire to plants breaks them down into their nutrients, and the ashes are used as fertilizer," Professor Muhajir said. "But ashes are very bad for the environment in the long term because when it rains, they leach out of the soil and pollute water. And there is also the carbon release from the plants. But people in villages don't understand this. They think burning is good."

"So, what are the solutions to prevent plant-burning? Are there any action steps suggested in this conference?"

"Well . . . it's complicated," uttered Professor Muhajir. The other men at the table nodded their agreement. More prodding from Sofia and me resulted in nothing beyond "It's complicated."

Here we were at this conference of environmental intellectuals, with thousands of years of combined experience between them, and they understood problems perfectly in all their intricate theoretical components, but they knew nothing of solutions. Their academic analysis seemed to have, in fact, moved them further away from solutions instead of bringing them closer.

The conversation was humbling and illuminating for me. It reminded me to keep my eye on solutions, to not conclude my journey with nothing in hand.

"*Ask for* I, NYOMAN KARI *School Teacher,*" commanded the colorful business card. Sofia and I asked the receptionist for him. Nyoman appeared immediately, as if by magic.

Nyoman was Indonesian but looked Indian, because his ancestors had migrated to Bali from India generations ago. He was bald except for wisps of carefully combed-over hair, and his face was unmemorable except for a wide, toothpaste-white smile. He was a high school teacher close to retirement, who'd married a former high school student. Sofia and I were staying at his bed-and-breakfast, which was a heavenly collection of a dozen quaint cottages set atop a luxurious valley of hundred-year-old trees vibrant with gliding birds and butterflies. It was among the most beautiful places we'd ever seen.

"You want come to cock-fight with me today?" Nyoman asked us.

Sofia and I looked at each other uncertainly. "You come find me at one o'clock," Nyoman told us, as if the matter were decided. Before he turned away to attend to other guests, he modified the time to two o'clock. And then to four o'clock. "Afternoon," he finally said.

We didn't bat an eye, nor did we reply. We knew by now that time in Indonesia is an imprecise and fluffy thing, like the clouds in the sky. A general time range—morning, afternoon, evening—is supposed to suffice for even the busiest of individuals, and an insistence on an exact time is as unseemly as an insistence on knowing someone's exact age. Using our best judgment about such matters, we decided to go find Nyoman at the middle time he mentioned: two o'clock.

Nyoman drove us to the cock-fighting ring in his jeep. Hundreds of men swarmed like bees around a circular arena with a dirt floor and a bamboo boundary. They pushed and pressed, spat and shoved, and smoked and drank. There was a frenzy in them—an insatiable bloodlust, a violent obsession; the scene was reminiscent of *The Gladiator.*

My presence and Sofia's created a smash of a stir. Heads turned, necks craned, and Nyoman became an instant celebrity in his village for having brought women to the cock-fight. Men came up to him with a grin and a handshake of felicitation, and they posed to him two questions in breathless succession: "Where are they from?" and "Are they twins?"

Two men stood in the cockpit, each with a rooster under his arm. One rooster was jet-black, with shimmering flourishes of teal-green, and the other was a subdued concoction of gold and bronze. The two men began pinching their roosters' throats and wattles, their fingers twisting in the flesh. They pulled the feathers at the back of the roosters' necks, and they slapped the top of their heads. The roosters began crowing in a piercing, grating way.

"They want to make it angry," Nyoman explained to Sofia and me, as we were cringing. "They do this every day for two years to make it angry."

A long, glinting, dagger-like spur was then roped to the left leg of each rooster. "Blade is very sharp!" Nyoman exclaimed. "It kill faster than beak!"

The first cock-fight of the day commenced and concluded in minutes. Sofia and I could see nothing. Several rows of men stood in front of us, forming a tight, unbreakable mass. Only when the cock-fight was over did we manage to sneak a peek over someone's shoulder.

The black-green rooster was on his feet but limping; he was the winner. The bronze-gold rooster was severely cut, bloody with the other's blade. His owner picked him up by a leg, flung him into a white sack, and tossed the sack aside, muttering angrily to himself. "He will make a soup out of it!" Nyoman laughed.

Two other cock-fighters with roosters entered the ring, and the same process was repeated with each combatant—chin-pinching, head-slapping, blade-tying, fighting. Sofia and I stayed for seven of the day's twenty-five cock-fights, all of them occurring one after another in an identical, agonizing stream of injuries and fatalities.

"What proportion of men in Bali like cockfighting?" I asked Nyoman.

"About 20 percent."

"What proportion of *them really* like it—are 'addicted' to it, as you say?"

"About 20 percent."

Twenty percent of 20 percent comes to 4 percent. If we exclude women, it comes to a mere 2 percent of the Balinese population. Only 2 percent of the people have a vested personal interest in cock-fighting, in other words. The tradition of cock-fighting continues, therefore, not because there are many people who love it, but because there are not enough people who oppose it.

It's the same way with farm animals, I realized. Only about 2 percent of Americans and Canadians work in agriculture, and it is primarily they who have a significant stake in any existing agricultural system, including hen cages and sow crates. These methods of confinement continue, therefore, not because there are many people who support them, but because there are not enough people who oppose them. There is no countering force to stand up to the very few who profit, by some measure, from confinement agriculture.

Cock-fighting has existed for thousands of years, but it is today morally and legally condemned in most developed countries. It is viewed as a barbaric ritual of the past on par with gladiator-fighting, bear-baiting, and witch-burning. In Indonesia even, cock-fighting has been banned for more than three decades and persists only through the bribery of local police.

Cruelty is not just cruelty, I realized also in the aftermath of the cock-fights. Institutional and individual cruelty are two different beasts. Cock-fighting is a form of individual cruelty; it is individuals who support it, and it is institutions—governments—who forbid it. Factory farming, in contrast, is a form of institutional cruelty; it is industries who endorse it, and it is individuals who, when they gain awareness, denounce it.

Individual cruelty toward farm animals decreases as nations become more civilized, but institutional cruelty increases, occurring now in the name of profit rather than pleasure.

After two days at Nyoman's homestay in Bali, Sofia and I took a five-hour ferry ride to a little island of sandy beaches called Gili Meno.

I'd never traveled by sea before, and I felt every upward and downward motion of the ferry in the pit of my stomach. But all benches were implanted with white, sun-tanning, sunglasses-wearing tourists, and so, in my dizzy, nauseated state, I saw no choice but to lie down on the floor like a dribbling drunkard. A crew member kindly gave me a mat. My eyes closed. I dozed.

"DOLPHINS!" a tourist screamed.

All the tourists stampeded to my side of the boat. One of them tripped over me, hardly stopping to apologize as she scrambled to remove her camera from its case. With her camera, she captured a photo of a fin visible over the waves. All this fuss about seeing *one* dolphin fin. The tourist-dolphin activity threw me over the edge. I threw up.

"It can only get better from here," I thought, breathing a sigh of relief. Before disembarking from the ferry, however, I threw up again. "It *has* to get better now," I told myself. But as I stepped onto crystalline white sands, I threw up yet again. I collapsed on the shore like spirit-broken Pi in *Life of Pi*.

As I gazed up at the stars, I realized that cock-fighting is like vomiting. *All* societal ills are like vomiting: they're obliterated only in steps. Banning a practice—whether cock-fighting or egg-laying hen cages—is merely a first step. Other necessary steps are societal awareness, acceptance, and involvement.

Sofia went to an island called Sulawesi to snorkel and dive, and I returned to Bali to study animal farming. I went to a Hindu neighborhood in the north of the island called Danau Bratan (*Daan-o-Brataan*), a spectacular, mystical-looking cluster of villages nestled in the mountains, where the clouds seemed to travel through the trees.

As I was exiting Danau Bratan's botanical gardens, I passed by a large sign: PICK YOUR OWN STRAWBERRIES. A man stood beside the sign with a yellow basket. He thrust the basket into my hands and opened the gate to his strawberry garden. Unwilling to offend—a pushover when it comes to such matters—I agreed to pick my own strawberries. But just when I started collecting them, the strawberry farmer trailing me like he feared I'd sneak some into my pocket, the clouds lashed us with a torrent of rain.

The farmer indicated a ramshackle shed to the side. We rushed in for shelter. He sat on a rickety chair, and I squatted on a footstool among soil tools and fertilizers. The rain pounded overhead, drumming a violent beat on the tin roof. The gardener's name was Agung. In only a matter of days, he would become among my closest friends in the world. In only a matter of weeks, he would be dead.

Agung was dressed like a Californian, in a white T-shirt, blue jeans, and baseball cap. He removed his baseball cap to dry it, revealing a smooth expanse with a shrub of hair around the perimeter. He smiled brightly, the smile adding childlike contours to a tanned countenance as round as the moon.

"Don't worry, be happy," he advised when he noticed me frowning at the rain. Those four words seemed to be Indonesia's motto, for people said them to me often. I hadn't realized before how frantically worried I looked. Agung, in contrast, exuded serenity and bliss, his face as unmarked as Buddha's. Later, he would share with me his secret: he meditated daily.

Agung had four tattoos, one on his chest, two on his right arm, and one on his left arm. All of them had their own significance, he explained. The bird on his chest represented freedom and grace. The sailboat-in-a-storm adorning one muscular arm represented the "crowded mind" he'd once possessed. The white woman's face tattooed above the sailboat represented Angelina Jolie, because he found her mesmerizing. His right arm was engraved with a roaring tiger, because "When I was in school, I was in a motorbike gang called Night Tigers." Agung also wore a sparkling, heart-shaped earring in one ear, because "Maybe I was woman in previous life."

A woman entered the shed. She was petite, dark-skinned, and strong-jawed, her shoulder-length black hair streaked rebelliously with auburn highlights. Her hands held two umbrellas and two cell phones. She was Agung's wife, Sri Kartika, and the only woman smoker I met in Indonesia, a fuming cigarette tucked eternally between her lips, smoking two packs a day to her husband's three.

Agung and Sri Kartika were both in their early forties and had met and married in their early thirties. They'd been trying to have children ever since, but were unable to. Agung found this ironic because "Her name means fertility!" His name meant "big," he continued, because he was the eldest of three brothers. He shared his name with the tallest mountain in Bali, called Mount Agung.

Agung spoke to me jovially in English, and Sri Kartika contributed to the conversation in Balinese, which he translated for me. The rain finally slowed, and we used Sri Kartika's umbrellas to cross the street to the restaurant that the two of them owned along with their strawberry garden. I spent the evening with Agung, Sri Kartika, and the half-dozen people who formed their "family, labor, and neighbor" classes, the three groups into which Agung laughingly divided everyone.

At the end of the night, under the pounding pellets of rain, Agung dropped me off on his motorbike to the bed-and-breakfast where I was staying. Called Strawberry Hill Hotel, it was a swarm of magically picturesque cottages ringed with ripe strawberries, rare flowers, and croaking frogs. The day I met Agung was my best in Indonesia.

I visited Agung's strawberry garden every day for the next week. He would pick me up from my cottage on his motorbike, and we would sit at a table outside his garden, talking for hours over baskets of strawberries, pears, apples, and a colorful Indonesian dessert that resembled muffins. We were joined often by Sri Kartika, family, labor, neighbors, or Doggy, Agung's dog who "sleeps in the yard, not like in America!"

Agung also often took me for a stroll in the village. We went to his cousin the bell-pepper farmer, his cousin the talented woodcarver, his cousin with the adorable (imported) golden retriever puppy. Agung and Sri Kartika also took me out for lunches and dinners, and they took me shopping to the local market, where they gifted me with dozens of pairs of earrings and accessories.

Agung was the kindest person I'd ever met. To him, kindness was not about sharing material wealth, but about being open, understanding, and accepting, sharing emotions and thoughts freely without expecting anything in return. Showered with smiles wherever he passed, he was liked and respected by one and all in his village. I was honored that he'd planted a foreign wallflower like me smack in the center of his thriving garden of family and friends.

"The first step to be spiritual is diet," Agung told me. "Before I become spiritual, I killed two pigs. I was sad when I killed them, and I had nightmare they try to kill me! After I become spiritual, I do not kill animals. I do not eat pig anymore. . . ."

It started raining as we sat outside the strawberry garden. It had rained every day of my stay in the village.

"Rain is because of the warming of the global," Agung informed me. "I have seen warming of the global in only the last two years of my life. It is raining for two years in this village, even in the dry season. This time of the year is supposed to be the dry season, but the wet and dry seasons are mixed now. I don't like this, because rain is bad for my strawberries; they need dry weather."

I was surprised to hear Agung talk about global warming, just as I was surprised throughout my stay in Indonesia to hear other villagers talk about it. "Do people know about global warming here?" I asked him.

"Yaa, of course. If you go to high school, you know it."

"Do you know that farm animals cause a lot of global warming?" I said. "They produce methane and carbon dioxide, and they cause more global warming than cars and all other world transportation combined."

"I do not know that," he said.

Most people don't. The terms "global warming" and "climate change" refer to a heating of the planet caused by increasing concentrations of greenhouse gases, which trap solar energy. Though some people are in denial about climate change, the overwhelming majority of scientists stand in concurrence that the climate is becoming more erratic, impacting rainfall and soil conditions. Animal agriculture, with its copious greenhouse gas emissions, is shown to contribute more to climate change than any other human activity. Climate change, in turn, affects food security, reducing crop yields of farmers like Agung because of shifting weather patterns.

With its high population density and biodiversity, and because it is an archipelago of islands, Indonesia is among the world's most vulnerable countries to climate change. In a 2014 speech in Jakarta, Indonesia's capital, U.S. Secretary of State John Kerry described climate change as "perhaps the world's most fearsome weapon of mass destruction." A three-foot rise in sea levels, he said, "would be enough to put half of Jakarta under water."

As I sat outside the strawberry garden with Agung, I realized that my journey into animal agriculture had thrust me into the worlds of so many different kinds of people, with whom I would have had no intersection otherwise. I now felt that previously, in my closed circle of Ivy-League, Wall-Street friends, I'd been missing out. *This* was real life, sitting outside a strawberry garden in conversation, drenched with rain. Many of my past life experiences seemed pale and colorless in comparison, like a dusty brown sparrow set beside a turbulently colored parrot.

My journey was also helping me find my emotional side. Previously, I'd been stoic, openly snickering and sneering at emotion. By tugging at my heartstrings, farm animals had shown me that I had heartstrings to begin with. Now that I'd opened the door to my heart, I wondered how I'd kept it closed so long.

I informed Agung that I would like to see some chicken farms. He told me that his younger brother Ketut (*Kay-toot*) would escort me to them.

In his early thirties, Ketut was surprisingly good-looking, with hazel eyes, broad shoulders, wavy brown hair, and a disarming smile. As all Balinese names have a meaning, most of them referring to birth order, so did Ketut's, meaning fourth-born. He was friendly but shy, which lent him a boyish charm. He spoke English more fluently than Agung, because he'd previously worked on a cruise ship, the dream job of many Indonesian men. Agung, in contrast, had previously worked at a hotel. Ketut spent the day escorting me to chicken farms on his motorbike, serving as my volunteer chauffeur and translator.

The first farm consisted of three barns, two of them two-story and one of them one-story. They were constructed of a wood floor and had net walls. I stepped inside the one-story barn, which contained 600 chickens. White-feathered, thirty-five days old, they would be sent to slaughter in two days.

A small area at the front of the barn was cordoned off from the rest with a metal barrier. This was the sick pen, it became evident, for the two dozen chickens within looked dead or close to dead. They all suffered from the same ailment: legs that were damaged as a result of extreme genetic selection for weight gain and rapid growth. Half of the sick pen chickens had legs that were no longer parallel to each other, but, rather, perpendicular—splayed out like a split.

One chicken, lying on his side, raised his head repeatedly to peck the floor in hunger. His desperate attempts were to no avail, however, for there was no food or water in the sick pen. These chickens were separated from the others, I realized then, not so that they could recuperate—their condition suggested little possibility for recuperation anyhow—but so that they could no longer *eat*. This way, no more money would be spent on them. Before dying, they would suffer not only the pain of broken legs but also the keen bite of starvation.

The other white chickens in the barn were not visibly lame, but almost all of them were sitting. On an impulse, I lunged toward them to gauge their reaction. They rose slowly . . . shifted a couple of feet . . . and plopped down again. They were so sedentary that they seemed almost sedated. I was struck with three simultaneous insights.

First, these uniformly white, industrially bred, imported, commoditized chickens were, in their appearance and behavior, almost an entirely different species from the cinnamon- and gold-colored local chickens who foraged in villages. White chickens were plump like beach balls and soft like pillows; local chickens, in contrast, were tall and lean, birds of build and beauty. White chickens sat still; local chickens sprinted, their legs so energetic and nimble that even snapping a photo of them was difficult.

Second, though pig farming and cow farming in Indonesia remain traditional, chicken farming is in a state of transition. It is shifting from local to white chickens and from small-holder backyard farms (flocks of one to two dozen chickens) to larger-scale agriculture (flocks of hundreds or thousands). The wealthier a village, the further along it is down the ratchet of agricultural industrialization, and, relatedly, meat consumption.

Finally, though this barn was similar in concept to chicken contract grower Terry's in Canada, Terry's was thirty-three times bigger, with 20,000 chickens per floor instead of 600. The number of animals does impact welfare at industrial farms. The white chickens I saw in Indonesia, albeit of the same painful breed, had a higher quality of life, living in an environment of less manure, cleaner air, more sunlight, and more space per animal.

On his motorbike, Ketut carted me next to an egg operation.

It was a shabby place bordered by a gray concrete wall, on which scurried a rat. When we opened the door, the first animals we saw were not hens but rats. They were everywhere—on the wood floor, in the shadowy corners, on the rusty ceiling beams—and they dispersed like pollen upon our entrance. I shuddered uncontrollably. "*Permisi*? *Permisi*?" Ketut said, meaning "excuse me," but there was no one to hear him.

There were six sets of hen cages in front of us. Each set consisted of four long horizontal rows, and each row consisted of 250 cages. Each cage was a small, square wood box that measured only about one foot in length, width, and height—and yet confined two hens.

The hens existed body on body, their necks and heads emerging like dying weeds from between the narrowly spaced slats in the roof of their cages. Most hens had bare necks, and it gave them a sickly appearance, like that of a patient undergoing chemotherapy. They'd lost their feathers by pecking one another and by scraping constantly against the wood slats of their cages. The hens were brown, and so were their eggs, as almost all are in Southeast Asia, where people culturally prefer brown eggs to white.

The egg-laying hens of Indonesia reminded me of Brick Roberts's egg-laying hens in Canada. The cages of Indonesia were made of wood instead of steel; they rose to two tiers instead of four; they confined two hens instead of four or five; and they were not as automated. But the concept was identical: incarceration for life.

Rats feasted on the mountains of dark, crumbling hen droppings underneath the cages. In cage-free systems, hens peck rats away, but in cages, hens are helpless, and rat colonies flourish. Rats are known to spread diseases like *Salmonella* and bird flu between hens at a particular farm, and also between hens at different farms, thus posing a sharp danger to animal and human health.

A woman emerged from a room to the side. Two young boys with wide, wary eyes stood behind her. The woman's name was Gede (*Gay-day*), which means first-born in Indonesia. Gede did not speak a word of English and I did not speak a word of Indonesian, but the language barrier was helpfully overcome by Ketut, who acted as translator.

Gede was kind, pleasant, and soft-spoken. She'd worked in this hen factory for nine long years, living in the small, cramped room behind her, which overflowed with rats. She'd given birth to her second son there. "We have twelve thousand hens here," she said, "and at least a thousand rats."

The hen cages had a shelf in front of them, onto which eggs rolled automatically. But eggs were not crated automatically: Gede collected them by hand every morning and placed them carefully in egg cartons. She held out her right arm for me to see, like she was a patient and I a doctor. The arm was covered from fingers to elbow with painful, bulbous pink scars. "The hens peck a lot," she said. "Every day, I find one cannibalized hen."

In front of us, one hen pecked the neck of her cagemate viciously, repeatedly. This was painful—especially given that the latter's neck was already a desert, sparse and featherless—but the cagemate was so accustomed and de-motivated that she did not react. Wincing at the sight, I held out my black hair-tie in front of the pecking hen to distract her from her cagemate. It worked; the hair-tie received a sharp peck.

Hens were aggressive because they were on the defensive. In their cages, they were unable to move away from their cagemates and Gede's arm—which descended daily on the egg belt—and so they sprung to attack. They were miserable, angry creatures.

"I don't think these hens would peck one another if they were outside cages," Gede told me through Ketut. "Local hens do not peck one another."

"Why is this cage system used in Indonesia?" I asked her. "Is it used to decrease the price of eggs or to increase the volume of eggs?"

"It is used to make my boss rich," she replied dryly. "My boss has five other egg farms. This village was hit with bird flu virus some years ago. My boss was the worst hit in the village. All his chickens were burned by the government to get rid of the flu."

"Were they burned alive or dead?" I asked her, fervently hoping that she'd reply "dead."

"Alive, of course."

The term "bird flu" is on the minds of all farmers around the world, spelling their worst nightmare. Viral diseases carry the potential to be exponentially more catastrophic in agriculture than bacterial ones, because they cannot be treated by dosing animals up with antibiotics. Viruses also often change and mutate rapidly, and jump species barriers, such that they are transmissible not only along human-to-human, but also along animal-to-human pathways.

A highly pathogenic H5N1 bird flu emerged in China in 2003 and spread to fifteen countries. It led to almost 400 laboratory-confirmed deaths (which tend to be a paltry fraction of total deaths). Because of Indonesia's vast size and island geography, it was the most affected country, suffering the highest numbers of both infections and deaths. Four out of every five Indonesians infected with H5N1 died. Gede and her sons could have been among them, for they didn't just work at the egg farm, but also lived there. The needle-sharp hen pecks on Gede's arm could have been a direct source of transmission.

A new strain of avian flu, H7N9, emerged in China in early 2013, ten years after H5N1. In its first fifteen months, the virus severely sickened 450 people and killed 165, again counting only laboratory-confirmed cases.

Notwithstanding the two avian flus of Asia, the worst flu of the last half-century has been the H1N1 swine flu pandemic of 2009, a potent assortment of human, pig, and avian viruses. It led the World Health Organization to declare its first-ever "public health emergency of international concern." H1N1 killed up to half a million people, according to estimates prepared by a team of more than thirty researchers. Though the virus emerged in Mexico, it spread rapidly to all parts of the world, infecting more than 200 countries in sum. Half the deaths occurred in Southeast Asia and Africa, the virus disproportionately affecting the poor. Unusually for viruses, which tend to target the elderly, H1N1 killed a disproportionate number of young people.

Another virus that targeted young people was the 1918 flu pandemic, which had chilling similarities with the 2009 pandemic. Called "the greatest medical holocaust in history" and "the mother of all pandemics," the 1918 flu is estimated to have infected one third of the world population—a total of 500 million people. It is conservatively estimated to have killed 50 million people, though the mortality could have been as high as 100 million.

Viruses can wreak unimaginable devastation, and conditions at factory farms are a breeding ground for them.

"This system of cages is banned in Europe," I told Gede and Ketut. "What do you think of that?"

"That is good!" Ketut exclaimed enthusiastically before Gede had a chance to reply.

Health and ethical issues relating to animal farms concern not only citizens of wealthy countries, but everyone—people like Gede more than most. I gave her money, feeling sorry for her and her young sons who, like the hens, lived trapped in a system in which gains accrued to owners and suffering was distributed among animals and workers.

Gede joined her hands together with gratitude in the traditional Balinese way, and she smiled with such pleasure that I wished I'd given her more.

Agung spoke often of his "big day," and I assumed this meant his birthday. It did not. It meant a ceremony day.

Bali is also called "The Island of the Gods," because everywhere on the island—in kitchens, restaurants, gardens, rice paddies—stand golden stone idols in the form of monkeys, tigers, dragons, and human-like deities, often wearing devious and enraged expressions. Ceremonies involving offerings of fruits and flowers are considered necessary to appease these volatile gods.

To my delight, Agung and Sri Kartika invited me to celebrate one of the most important Balinese ceremonies of the year with them and their family at the nearby lake temple. There was one problem, though: tourists are forbidden from entering the prayer sections of the temple. Sri Kartika overcame the challenge by taking care to dress me elaborately in her own festive ceremony clothes, promising "No one will know you are not Indonesian."

I wore a deep-blue sarong with a golden hem, along with a long white shirt embroidered with delicate gray flowers. Sri Kartika tied my hair up like hers, and Agung inserted a lovely lime-green orchid from his garden into it.

The lake temple was a collection of tall, beautiful pagoda-style thatches on a bed of shimmering blue water hemmed with mountains. As I kneeled on the floor of the temple, sitting between Ketut and Sri Kartika, and as I was sprayed with holy water and sprinkled with rice by Hindu priests, I not only looked, but also felt, truly and happily Indonesian.

The feeling increased when, after we returned to their house, Agung showed me his shrine. It was his spiritual retreat, where he came to meditate and contemplate daily. He presented me with a small golden and ivory statue of a Hindu deity called Shiva, who was meditating cross-legged, one hand clasped to his knee, the other hand raised to his heart. Shiva, in his powerful pose, reflected

Agung's personal transformation from a man with a "crowded mind" to the embodiment of calm. Agung hoped I would experience a similar transformation; so did I. I was honored to receive such a thoughtful and meaningful gift from him.

"I am sad I may never see you again," Agung told me. "Promise me you will return to Bali for your honeymoon."

I had no prospects of marriage, but I imagined (hoped) I would someday, and I made my promise, intending to keep it.

On my last day in Indonesia, Agung, Sri Kartika, and Ketut kindly drove me two and a half hours away from their village to the airport. I wrote down Agung's and Sri Kartika's birthdays—only six days apart in December—so that I could call them every year and send them a gift. The two of them had showered me with affection, conversation, and presents, and I was eager to start returning their thoughtful gestures. But I never got a chance. Just weeks after my departure, I received a note from Ketut that Agung had died.

Agung's chest and back broke into spasms of sudden, unprecedented pain. Ketut rushed him to the hospital. Agung appeared to feel better in a matter of hours, rising to play cards through the night, as he enjoyed doing. But he collapsed suddenly in the early morning. His priest attributed the death to black magic. The true medical cause was unknown.

I was devastated at Agung's loss. I cried for days.

Our lives had crossed only briefly, but in that short period Agung had become my friend, my role model, my mentor—my *guru*. He'd taught me the importance of happiness and the meaning of kindness. He'd showed me to listen to my heart rather than neglect it. He'd inspired me to become a better person, and to, in turn, create a better world.

I simply could not believe he was gone. I was heartbroken.

CHAPTER TEN

FAST FOOD NATION MALAYSIA

A NICE BIG AC HOUSE FOR CHICKENS

As the airplane descended, I did not see people. I saw palm trees. They stood to either side of the roads like giant pineapples, like leafy umbrellas. Their thin-fingered branches swayed in the breeze, and their trunks stood still, planting the entire country of Malaysia firmly into the world economy. Palm oil has, to Malaysia, been liquid gold, a waterfall that never runs dry, comparable to the oil and gas fields of the Middle East. Malaysia is the world's largest exporter of palm oil.

"Lady, to me, palm plantations are good, bad, and ugly," the airport taxi driver informed me as we sped past landslide warning signs on roads. "Good for economy, bad for environment, ugly for eyes."

From high up in the air, Malaysia looks little like its southern neighbor Indonesia, and the difference is even starker from the ground. Indonesia is a developing country, Malaysia a developed one. Indonesia is proud of its wide-ranging personality, Malaysia strives for hard-nosed productivity. Their national mottos hint at their differences: Indonesia's is "Unity in Diversity" and Malaysia's is "Unity is Strength." The houses in Malaysia are bigger than in Indonesia, the buildings are taller, the roads cleaner, and the mosques and museums more impressive in architecture.

Malaysia and Indonesia are similar in their lushly tropical climate—moist like a teardrop in summer months—and they rank equally well in literacy and life expectancy. But they are different in most other demographic respects. Malaysia's population of 30 million is one eighth of Indonesia's, and the two countries are comparable in their absolute and relative population size to Canada and the United States, respectively. In Malaysia, over half the population is ethnically Malay, close to a quarter is Chinese, and 7 percent is Indian. Malaysia's national language is Malay (which is similar to Indonesian), but the language of school and business is English (an advantage to me), because the country was once a British colony.

Malaysian gross domestic product (GDP) per capita is triple Indonesia's, and shopping and dining form the nation's primary excursions. Malaysian malls are grand and air-conditioned, brimming with high-end American and European stores. Supermarkets are also fast replacing outdoor markets. Rows of cow heads with eerily alert eyes, stacks of yellow chicken claws pointed to the sky, buckets of floppy, pink pig ears in pairs, are giving way to meats packaged in transparent plastics, the body parts and the animals they originate from increasingly indistinguishable from one another.

The most popular lunch and dinner options in the country are Kentucky Fried Chicken (KFC), McDonald's, and Pizza Hut.

Malaysia is home to more than 500 KFCs, 300 Pizza Huts, and 300 McDonald's, which together serve millions of customers every week. A global survey in 2004 found that Malaysians eat even more fast food than Americans. One third of American adults eat fast food at least once a week, in comparison to a startling three fifths of Malaysians, second in their hamburger worship only to the people of Hong Kong.

Malaysia's booming fast food sector is part of the reason for its expanding waistlines. One out of every seven Malaysians today is obese, in comparison to one out of every twenty Indonesians. The number is still low when compared to the United States, where one third of adults are obese and another one third overweight.

Malaysians are united in their approval of their nation. They enjoy a comfortable standard of living in a growing economy under a stable government. They view their immense neighbor Indonesia as a pitiable place for its overpopulation, and they view their minuscule neighbor Singapore as a dot whose citizens work like robots. Malaysians believe they have achieved just the right balance, and are on their way to just the right future.

MARDI in Malaysia refers not to the rowdy "Mardi gras" celebration that occurs in New Orleans every year, but to the Malaysian Agricultural Research and Development Institute. Everybody told me to go MARDI, so I did. There, I spoke to Dr. Shan (*Shaan*) for two hours.

Dr. Shan was a tall, bespectacled Indian man with a thick gray moustache. He was friendly and fatherly, having two children close to my age, and he was professional and personable, smiling throughout our conversation. Dr. Shan had obtained a veterinary degree and Ph.D. in the United Kingdom. He was the "Deputy Director Production System" of a MARDI division known as the

Strategic Livestock Research Centre. Of his two decades in the division, the first had focused on chickens and the second on pigs.

"Are you a student?" he asked me.

"Yes. I'm studying food production."

"Do you have a focus?"

"Yes, international agriculture. Animal farms specifically."

"Is it a master's or a Ph.D.?"

"Master's."

As soon as I weaved my web of deception, I wished I hadn't. Dr. Shan was kind, knowledgeable, and respectable—the sort of person I instinctively wanted to impress. I felt awful lying to him. I felt so bad, in fact, that I would call him before leaving Malaysia and come within a hair's breadth of telling him the truth: that I was not pursuing an official degree, but my own haphazard curiosity, which I myself was at a loss to understand. How could I explain it to him when I could not understand it myself?

"In Malaysia, there used to be village backyard farming like there is in Indonesia," Dr. Shan said, commencing his verbal tour of Malaysian agriculture. "But now it is completely gone. All poultry farming here is industrial, both for broiler [meat] chickens and layers [egg-laying hens]. There are two kinds of houses for chickens in Malaysia. They are called open houses or closed houses. Open houses have netting instead of walls, and they have natural light. They tend to have about six thousand broiler chickens."

Open houses were the kinds of farms I'd seen in Indonesia, except that those had been one-tenth the size of their Malaysian counterparts, with 600 chickens instead of 6,000.

"Closed houses are much bigger than open houses," Dr. Shan continued. "They are called 'closed' houses because they are kept closed. The only way for outside air or light to enter is through exhaust fans. Closed houses usually have twenty thousand chickens in one house. Or, if the house has two floors, they have forty thousand chickens. There are usually five houses on one site. That means that one site by itself can have one hundred thousand to two hundred thousand chickens at a time, and in one year, the total production can be a million chickens."

Closed houses were the kind of operation I'd seen in Canada. Terry's broiler chicken farm had precisely 20,000 chickens on each of its two floors. The number 20,000, I was learning, was the industry's standard number of chickens per floor anywhere in the world.

"Malaysians are eating more and more meat, especially chicken," Dr. Shan continued. "It is because of the social factor. When people get rich, they put their money in meat. When I was ten or twelve, I ate chicken once or twice a year. Now, I eat it once or twice a day. Chicken was once a luxury everywhere in the world—it was more expensive than beef—but today we can eat as much of it as we want."

In Indonesia, the staple meat continues to be beef, but in Malaysia, it has become chicken. Chicken consumption is one factor that shows how firmly Malaysia has joined the ranks of developed nations; chicken has been gaining steady popularity in the west for decades, as beef has simultaneously been shrinking on the plate.

Beef's share in world meat consumption has plummeted from 40 percent a half-century ago to less than 25 percent today, due primarily to health concerns. In contrast, chicken's share in world meat consumption has more than doubled from 15 percent in 1970 to 32 percent today. Pork is the most popular meat on the planet today. Most expansion in both pig production and consumption in the last half-century, however, can be attributed to China, which consumes close to half of the world's pigs. Chicken is expected to remain the fastest-growing animal sector in the future, expanding significantly even in developed countries, in contrast to beef and pork.

"Industrial farming started in Malaysia much later than in the West," said Dr. Shan. "It started here in the 1990s, when big companies started coming in as 'integrators.' The word 'integrator' means that every step of the production process is integrated. The integrator gives contract growers day-old chicks to grow, then it buys them back for slaughter. The majority of poultry in Malaysia comes from contract growers linked to integrators.

"Here, it's exactly the same kind of system as Tyson, the big chicken company in the States. All the farm technology here is adopted from overseas. Breeds are from abroad, equipment is from abroad. Techniques are also from abroad—we are starting to use antibiotics and artificial insemination here, too. Today, there is little difference between American and Malaysian systems and productivity.

"Integrators were not popular here at first, because people were skeptical. They didn't understand that agriculture could work like this—in closed houses, in high numbers. But then we—MARDI, the government—convinced them that this is the only way forward. We told people that closed houses are more biosecure, more productive, and also more comfortable for chickens. Chickens in closed houses have stress-free lives. They have better lives than we do."

"Do you really think so? That they have better lives than we do?"

Dr. Shan laughed sheepishly. "Maybe not," he admitted. "But there is no one to say otherwise."

The official factory-farm propaganda had been given to the country, and, uncontested, it had been accepted by the country. It had been accepted by producers and consumers, by young and old, by educated and uneducated. In fact, the taxi driver who'd driven me to MARDI had enthusiastically informed me: "In Malaysia, they make a nice big AC [air-conditioned] house for chickens. It is heaven for them."

The battle today in animal agriculture is not just about production systems, but also about animals themselves, about their very nature. Agribusiness is on a rampage to convince itself and everyone else that sunlight, clean air, and space to move are not necessities for animals, nor even luxuries, but, rather, acute forms of deprivation. As such, discussion is perverted before it can even begin.

"How can I visit Malaysian farms?" I asked Dr. Shan.

"You can't. Biosecurity is very strict here. You will have to make a special request to get in—you will have to write a letter. You will also have to shower. Farms don't want people to enter. I work for the government, but even for me, it is difficult to enter."

"You have letter?" asked Mr. Hubib (*Hub-eeb*). "Letter from your university?"

"No."

Mr. Hubib's eyes narrowed as he peered into my face. Detecting nothing in my carefully constructed blankness, he consented, "Okay, fine, I show you my farm fast."

Mr. Hubib was a chubby, fair-skinned Malay man who looked Chinese. (I called him "Mr." because in Malaysia, a country of courtesies, a title is required for anyone older than oneself.) Mr. Hubib wore a yellow sports jersey, and a cell phone dangled from a chain around his neck like a pendant. He was a father of three boys, one of whom was a baby, gauging by the English BABY IN CAR sign on the rearview window of his blue car.

"I work in petrochemical plant before," Mr. Hubib said. "Then I come here because my brother start this chicken farm to make some money. It is more tolerable to work with family than with other people. I am contract farmer for KFC. KFC have many, many farm in Malaysia. KFC guide us on what to do. They give us DOC, then they take back DOC for harvest."

Mr. Hubib's vocabulary was illuminating. To him, it was not a "day-old chick," but a "DOC," and it was not slaughter, but harvest—a euphemism that suggested that these were not animals, but plants. Our thoughts help us choose our words, but our words also help us choose our thoughts.

"Many of the chicken in Malaysia are for KFC," Mr. Hubib continued. "Nowaday, most of the people like KFC. In KFC, you can have your lunch, you can have your dinner, you can have your breakfast. It's very fast, you see. It is called *fast* food. All people can eat in KFC—Muslim, Hindu, Chinese—because KFC have no pig and no beef. Okay for everyone. KFC also have very good taste and very good price."

KFC is the fast food leader in Malaysia, for four reasons. First, as Mr. Hubib said, chickens are culturally the safest animal to eat, offending no religious sensibilities, in contrast to cows and pigs. Second, KFC's entry to Malaysia precedes that of McDonald's by a few years—but a crucial few years. KFC arrived in 1973, seven years before McDonald's in 1980, giving Malaysians more time to develop a taste for fried chicken than for hamburgers.

Third, KFC spends millions of dollars every year on innovative marketing and branding. In 2007, KFC Malaysia launched as its logo the friendly, elderly Colonel Sanders face that is ubiquitous in the United States. The launch was successful, "creating excitement at the restaurants" and "improving throughput." Finally, not only has KFC mastered promotion, but it exercises iron control over production, as Mr. Hubib explained to me.

"KFC have different model than McDonald," Mr. Hubib said. "McDonald buy meat from companies; KFC have own full chain, from chick to slaughter to tray. KFC own everything."

KFC ranks as one of the top five integrators in Malaysia, hatching, growing, and slaughtering more than three million chickens every month. In comparison, McDonald's level of integration in the country is far less advanced and established.

Mr. Hubib was one of more than a hundred KFC contract growers in the country. He led me to one of his four chicken warehouses.

There were 20,000 chickens in front of us, and 20,000 on the floor above us, exactly as Dr. Shan of MARDI had described. Mr. Hubib's site produced a total of 160,000 chickens every month.

Outside, the weather was bright and sunny, but inside, it was blue—literally. The walls were covered with a dark blue canvas that made the place look space-age, and had an oddly tranquilizing effect. "Blue make the chicken calm," Mr. Hubib explained.

He was scientifically right. Blue has a calming effect even in humans, but birds are more sensitive to color than people. A *British Poultry Science* article explains that the color blue also "accelerates muscle and breast growth, increases body weight, and may create softer, more tender muscles." When it comes to profitability, agribusiness is extremely attuned to animal biology.

Mr. Hubib's sixteen-day-old chickens did look calm—comatose. I reached my hands into their midst and snatched up two of them. They were so lethargic that they didn't even try to get away. Their bellies were plump, pink, and bare of feathers, and their wings were lumpy stumps. Their large feet flailed through the air comically as they shrieked in my hands. They were ghostly caricatures of chickens—harmless, but terrifying to witness in their harmlessness.

Although I'd seen tens of thousands of broiler chickens by now, I still could not get over how fast they grew. I found it mind-boggling to imagine that, were I to return three weeks later, they would already be slaughtered, having reached the requisite weight in a total of only five weeks.

Mr. Hubib, who'd been a KFC contract grower for five years, was also continually confounded by the chickens' growth. "Chicken grow very fast," he said. "Breed is name Cobb. I know name is Cobb because it is on form KFC give me. Cobb is chicken breed of Tyson in America."

It struck me that in addition to Mr. Hubib, Dr. Shan had also mentioned Tyson. Most Americans do not know the name Tyson—they do not know that it is the largest chicken producer in their country and in the world—but here, in Malaysian agribusiness, the name Tyson is thrown around like that of a next-door neighbor.

In the United States alone, Tyson slaughters more than two billion chickens a year, along with twenty million pigs and seven million cattle. "Nearly all Americans regularly eat Tyson meat—at home, at McDonald's, at a cafeteria, at a nursing home," writes Nicholas Kristof in *The New York Times*. By working with thousands of contract growers, Tyson has invented, perfected, and exported the concepts of contract growing and integration.

Tyson is so influential that it *owns* a breed of chicken, called Cobb, which it began experimentally developing in the 1980s. From the beginning, what has set Cobb chickens apart from others is their oversized breast. The breast's size is purposeful, as Tyson believes that "broilers with the most breast meat offer the greatest return to the producer" and look better when "tray-packed under clear plastic." The large breast has resulted in Cobb chickens often being unable to stand up straight, instead plopping forward onto their breast, frequently on collapsed legs. Three out of four chickens in Malaysia today are Cobb, originating directly from Tyson.

"Sometime we feel that DOC is not a good, healthy DOC," Mr. Hubib lamented. "Look on movement of chicken. Healthy chicken always moving." He moved two fingers rapidly to demonstrate.

"Cobb chicken not moving. Cobb chicken sitting. And when sitting, spreading legs behind and putting head down. This not natural. In my experience, bird putting head down when sick. This mean, Cobb chicken sick."

In fact, this was the precise state of many of Mr. Hubib's chickens. They'd buried their heads in the rice husks on the floor, their eyes closed, and they'd spread their legs behind them, such that their entire body was horizontal, breast-down, like an elephant seal's. Their breast, that elusive profit-maker, was too heavy for them to hold up.

"And these chicken aggressive," Mr. Hubib went on. "Outside local chicken play with friends, but don't draw blood or peck on it. Here, I experience biting of friends until injured."

Initially, I thought Mr. Hubib was talking about his friends nibbling on him, but then I understood that he was saying that his chickens sometimes pecked one another to the point of drawing blood.

Mr. Hubib checked the time on the cell phone dangling from the chain around his neck. "It is one o'clock, and it is Friday!" he exclaimed. "I go pray now at mosque. Weekly prayer time."

Mr. Hubib and I exited his KFC, Cobb-chicken warehouse. My eyes teared from the effort it took to adjust from the hypnotic blue hues inside to the sunshine. But my relief at being out of the blue was short-lived. Mr. Hubib dropped a bomb on me.

"I have open house before," Mr. Hubib said, "but in open house, you have to give 1.0 to 1.2 square foot of space per chicken. In closed house like I show you, you can have more chicken, and you give chicken only 0.75 square foot per chicken. Now, I have new kind of farm in addition to this closed-house farm you see!"

His round face lit up like a lantern. "My new farm have four barn that are more advanced than these four barn. That farm use system designed for egg-layer chicken!"

"What are you talking about?" I asked, not wanting to know.

"My new broiler-chicken farm for KFC have cages."

"But I thought cages are only for egg-laying hens," I said, "not for chickens raised for meat consumption?"

"Yes, before, but not anymore! Broiler cages start in Malaysia last year. Big development. In broiler cage, I give chicken only 0.44 square foot, and I have *lot* of chicken. I have fifty-four thousand chicken in one barn! I make lot of money."

I was floored. I knew of cage setups for egg-laying hens, but I'd never heard of cages for broiler chickens before. I was hoping that I'd misunderstood Mr. Hubib's confusing English. I would have to see the broiler cages to believe them. Mr. Hubib, however, didn't like the idea, wincing as he clutched the cell phone around his neck with both hands.

"*Haiyya a'laa salah*!" he exclaimed. The Arabic words mean: come to prayer.

"My cage farm is two kilometer away, for biosecurity," he said impatiently. "No time today. I go to mosque to pray now."

He gave his chickens 0.44 square foot of space each—about the size of this book—and he thought prayer would save his soul?

Mr. Hubib eventually conceded, appearing motivated by the idea of showing off his new facilities. "I give you fifteen minute more," he said with a sigh.

Mr. Hubib's new site consisted of four warehouses in a row, as before. As he and I walked toward the second of them from his blue car, my legs kept pace with his, but they seemed wooden. I was tired of it all, I realized wearily. I felt exhausted and overwhelmed, like a cell phone whose battery was close to dying. I didn't want to see any more. I didn't want to know any more. I wanted to run away and never return.

Mr. Hubib and I entered. The place stank powerfully of ammonia. There were three rows of metal cages before us, the highest I'd ever seen. They rose intimidatingly to five tiers—more than eight feet tall in all—such that my neck had to crane to glance at the highest tier. Each cage was about the length and width of a desk and held 120 chickens. The chickens were so crowded that it was difficult to tell where one ended and the other began. They disappeared into one another like white clouds—an intermingling mass instead of distinct individuals. In each cage, all but one or two chickens were plopped down.

Mr. Hubib's other farm had been blue; this one was black, immersed in a cave-like darkness. Black was so much worse than blue. Brick Roberts's cages in Canada had been dark, too, but not as dark as these, because when it comes to egg-laying hens, pecks do not directly impact profitability. When it comes to broiler chickens, they do.

"If chicken peck friend, peck leave mark on skin," Mr. Hubib explained. "This is bad for price of meat in market. People don't like to see mark on skin when they buy. We close light so chicken cannot see and chicken cannot peck."

Consistent texture of meat in the grocery store requires consistent torture at the farm.

Broiler chickens have not been caged until recently because of fears of breast blisters and scratches against cage wire. Technological development has taken care of these fears today: instead of sitting on a wire floor, Mr. Hubib's chickens sat on a yellow, plastic-covered mesh that prevented abrasions. Plastic flooring protected the chickens from wire, and darkness protected them from one another. Protection and deprivation worked hand in hand.

Mr. Hubib went off to the mosque to say his prayers, and I went off to take a boat ride across a river.

The water was still and stunning, cool to the dipping of my grimy fingertips. The sun was warm on my arms and the sky was bright blue, streaked with wispy clouds. Trees edged the river, looking like broccoli florets from afar, their green reflections floating like nets upon the water. Buffalo grazed along the shore, their tails swinging lazily in the breeze.

The boat ride was indulgent but essential—a massage for my strained mind. I finally admitted it to myself on the river: I was getting depressed.

My mind was a mess. My sleep was sleepless. I was a world away from everyone I knew. At least in Indonesia, I'd had my sister; in Malaysia, I had no one. My quest to solve the problems of animal agriculture seemed as hopeless and inexplicable as a dog chasing his own tail. I'd dived too deep into the swirling sea of the industry, and I had no more energy left to continue swimming, to continue thrusting one arm before the other against the tide and power of the water. No matter where I turned, more waves crashed onto my head, sinking and suffocating me under the surface.

The boat ride helped for the time being. As I looked out at the water, I realized that such are the times—midsummers and middays—when life is most sensory, most beautiful, most worth living, for man and animal alike. But farm animals today are deprived of even such basic pleasures, such "creature comforts." They are confined, caged, and controlled, never once seeing sun, sky, or grass. But to what end?

Floor tiles shone like mirrors as they reflected ceiling lights. Paintings with ornate frames depicted galloping horses and swishing waterfalls. The front desk was adorned with a pastoral statue of a hen resting on a nest, surrounded by a brood of chicks. The sofa on which I waited had plush blue cushioning and an extravagant golden border. The headquarters of Huat Lai (*Huaatt Laaey*) were indulgent.

Behind the glass walls to either side of the lobby stood rows of offices and cubicles. Men and women rushed between them with a sense of urgent importance, attired in pressed shirts and tailored pants.

The two scarf-wearing receptionists at the front desk were about my age but appeared much younger in their demeanor, their expressions bright and trusting in comparison to my old, worn look. I told the receptionists I was interested in seeing their company's egg farms. They shook their heads at me in unison, giggling.

Part of the reason for their refusal, I knew, was my attire. I was dressed like a homeless person. This was despite the fact that I was that day wearing the best outfit I'd brought with me to Asia: a crumpled beige skirt, a faded violet sweater, and stained pink flip-flops. I'd packed just one small, hand-carry suitcase for my entire Asian expedition, figuring it would make it easier for me to embark on planes, buses, and taxis without hassle. The trouble with my brilliant plan was that I had little to wear.

As I watched the self-important employees behind the glass walls, I thought with nostalgia of the ironed shirts and dry-cleaned suits I'd worn on Wall Street. Now, as I looked at myself through the receptionists' eyes—my old clothes, my tangled hair, my frowning forehead—I felt embarrassed at my unkempt state.

Feeling like I was bullying children, I bullied the receptionists into making phone calls to their superiors. As a result of their efforts, a tall, slim woman—who reminded me irritatingly of my former self—came out to see me. "You need an official letter from your university if you want a tour," she informed me before turning on her heels and hurrying back to her desk.

I was turning around to leave, when a Mr. Lim dropped by the front desk. He was Chinese in origin but gave the two receptionists business orders in Malay. (Most people in Southeast Asia speak multiple languages fluently.) As the receptionists made phone calls and he waited, he seemed to feel forced by propriety—a vital part of Malaysian culture—to commence a conversation with me. "Come alone?" he asked with little interest.

"Yes, I came to Malaysia alone."

Mr. Lim turned to face me fully, his jaw dropping, his eyes wide. "Come *ALONE*?!" he repeated. "*Completely* alone? Not scared?"

"Yes, I came to Malaysia completely alone. No, I am not scared."

Everyone I'd met in Malaysia so far was shocked that I was traveling alone. "How enterprising of you," Dr. Shan had said. "How daring you are," a cab driver had said. "I have never heard of that before!" a hotel clerk had exclaimed. "Most lady are scared," a restaurant owner had said.

But no one was as shaken as Mr. Lim. "Come *alone*. . ." he repeated weakly, his face pale and slack. He looked as rattled as if I'd told him I'd invented the moon. Neglecting the business that had brought him to the front desk, he plodded grimly toward a meeting room. He was too frazzled to tell me to follow him, but I sensed I should.

Mr. Lim seated himself on a roomy, black leather chair at the head of a long, polished boardroom table. I'd assumed when I'd first seen him that he was a midlevel employee; his demeanor and instinctive possession of the main chair suggested I was wrong.

I found Mr. Lim's face remarkable in how unremarkable it was. If I were to run into him on the street an hour later, I would fail to recognize him. Each of his features was arranged succinctly with all the others, as if organized by the law of efficiency, none taking more space than it should. Mr. Lim's leg jerked impatiently under the table, his toes tapping on the soft carpet. Though his face held still, it showed the same background stress as his incessant leg. I imagined that he was probably calculating the setback to his schedule caused by his impromptu meeting with me. He seemed the epitome of productivity.

Mr. Lim, fifty-two years old, was today at the acme of achievement, despite his humble beginnings. He'd begun his agricultural career in 1979 as a lowly worker at a small egg-laying hen farm. A decade later, he'd left to start his own egg operation of 25,000 egg-laying hens. That one facility, in a matter of years, had given birth to Huat Lai, the vast, integrated powerhouse that Mr. Lim directed today.

Huat Lai was among the top five poultry corporations in Malaysia, owning more than two million egg-laying hens and two million broiler chickens. It consisted of an umbrella of nine subsidiaries, each managed by one of Mr. Lim's nine siblings—eight brothers and one sister. Sub-managers of the subsidiaries formed a hefty class of uncles and male cousins. Huat Lai was today a public company trading on the stock exchange, but it was still majority-owned by Mr. Lim and his siblings, and was thus technically still a "family" operation—a fact that showed me, yet again, how little the term means.

Malaysian agribusiness, like American agribusiness, is an oligopoly. Only a handful of corporations supply the majority of the country's chickens and eggs. Huat Lai managed some of its chicken and egg factory farms itself, and others were managed by its contract growers. All chickens were slaughtered in Huat Lai's own slaughter plant. The company was fully integrated—it even made its own egg-carton trays.

Most of Huat Lai's chickens and eggs were sold in Malaysia, and others were exported to Singapore, China, Hong Kong, and the Middle East. Malaysian customers included large, international supermarket chains like Carrefour, Tesco, and Giant. Huat Lai's largest customer was McDonald's.

Mr. Lim and I got into a car with a chauffeur, who drove us to Huat Lai's slaughter plant a few minutes away.

As we strode through the slaughter plant's vast lobby, we passed an office that belonged to Mr. Lim's brother, the one in charge of the slaughter subsidiary. Though I caught only a glimpse of the office through the door left ajar, it was enough for me to know that the office was the most luxurious I'd ever seen. It had a fully stocked alcohol bar shaped like a quarter-circle, its desk was imposing and gleaming, and its carpet was lushly opulent. If this was his brother's office, I wondered how Mr. Lim's office might look.

The office smelled strongly of air freshener. The reason for the air freshener became apparent when Mr. Lim opened the next door, onto the landing above the stinking kill floor.

There were five long windows on the landing, each of them giving a view onto one section of the kill floor. More than a hundred workers, almost all of them male, worked in identical white clothes, performing their assigned tasks without speaking, without glancing right or left—cogs in a giant killing complex. The kill floor was a stark sight from above, a high-tech, space-age vision from the future, like a scene from Aldous Huxley's *Brave New World*.

Through the first window, workers could be seen unloading chickens from red crates onto conveyor belts, hurling them like basketballs, as if the chickens were devoid of physical sensation and the workers of psychological sensation. The second window should have shown the slitting of chicken throats, but a barrier was erected on the kill floor such that the slaughter could not be observed from above. The executives and customers of Huat Lai—for whose purpose the windows were carved overhead to begin with—wished to eschew the sight of killing.

Through the third window, dead chickens could be seen zipping by on a line. Along the way, their heads were snapped off by one machine and their feet by another. An eviscerator then reached inside them and clawed out their intestines. The line of eviscerated chickens raced next into the chill room, visible through window four, where it ran in circles for a total distance of a mile, in order to cool the flesh faster.

The final window beheld the most activity. Workers stood to either side of long conveyor belts of chickens, hacking off bones and bits with knives. Machines automatically packed chickens in transparent plastic wraps, and workers stacked the packages in crates that would later take them to McDonald's and other restaurant and retail chains. The chickens had entered the kill floor in crates, and they left in crates, but at the end, no one would be able to argue that the pale pink flesh wrapped in plastic had ever belonged to a living animal. The idea seemed ludicrous even to me.

I asked Mr. Lim if I could see his farms, too.

He was silent for several seconds, and I wondered if he hadn't heard me. "Okay," he eventually replied. "My people will take you to farm tomorrow. Come at nine."

At nine o'clock sharp the next morning, Mr. Lim's "people" were waiting for me at headquarters.

One of them was a pudgy, friendly Malay chauffeur and the other was my assigned farm guide, Mr. Tan. A Chinese man in his thirties, he had the same air of number-crunching efficiency as Mr. Lim. He was trimly dressed and his fingernails were trimly cut, except for a deliberately long pinky nail. He seemed politely resigned to the duty of showing me around, but also keen to dispense with me swiftly.

"You only one people?" Mr. Tan asked me, in the controlled tone of verifying something he'd heard but couldn't convince himself to believe. Upon my affirmative response, he voiced "*Wow.*" The driver shook his head disapprovingly with *tut-tut* noises.

"How you know my boss?" Mr. Tan asked me.

"I talked to him for a while yesterday," I replied.

Mr. Tan straightened in his seat, looking impressed.

"We take you to biggest of our four layer sites," he said. "That site have twenty houses for layers [egg-laying hens]."

When we arrived at the site, we stepped first into the egg-packing facility. With its high, sloping ceilings, it was the place where eggs from the twenty surrounding hen warehouses were washed, graded, packed, and stacked. Forty or so workers milled about, attired in a uniform of yellow company shirts and navy pants. Half of them were Malay and the other half were imported from poor countries like Indonesia, India, Nepal, and Bangladesh. One section of the egg-packing facility contained stacks of large cardboard cartons full of eggs intended for Hong Kong. The cartons had the word "fresh" written on them in English.

Food companies everywhere, I realized then, know that consumers everywhere like the word "fresh." But what does the word "fresh" even mean on eggs? What exactly *are* "fresh" eggs? Are they eggs that were laid recently, thus experiencing a short time to market between egg-lay and grocery shelf stocking? If so, how can eggs transported across international borders be called "fresh"? Or are "fresh" eggs those laid by hens who have lived in natural conditions, having, for instance, breathed "fresh" air? In that case, how can factory-farm eggs be fresh? The word "fresh" is a marketing gimmick that has no place in agribusiness.

Mr. Tan handed me a white Huat Lai shirt, a pair of navy blue pants, a towel, a bar of soap, and a white mask. Fortunately, the pants had an elastic band that I could tighten to prevent an accident of the kind that had occurred at the Blackwater slaughter plant in Canada. Of all things, I was by now experienced in the apparel of agribusiness. I *knew* how it worked—the showers, the clothes—and it didn't faze me anymore.

When Mr. Tan and I reunited in the corridor after our biosecurity showers, we were dressed identically. There was just one difference between us: Mr. Tan wore his white mask over his nose and mouth; I didn't, because I thought it looked silly.

The twenty hen facilities were very long and very gray, each the length of two to three football fields. We entered the third of them, which Mr. Tan said held the healthiest hens.

It didn't look like it. In front of us stood five rows of metal cages stacked up five levels, each cage confining seven egg-laying hens. In total, there were 65,000 egg-laying hens under one roof—the most I'd ever seen—a full *five* times as many as I'd seen at egg-laying hen operations in Canada and Indonesia. Each hen at Huat Lai was accorded one third of a square foot of space in which to get on with her existence.

"This is our old design of farm," Mr. Tan told me. "This house is eleven years old. In our new design, we will have cages of seven floors, not five. We will have ninety thousand hens per house."

I couldn't even imagine how tall seven-floor cages would be. I didn't want to imagine it.

Huat Lai hens were auburn in color and belonged to an industrial, internationally exported German breed called Lohmann. Huat Lai cages were imported from Italy, from a large corporation called Tecno that sells its cages in more than 90 percent of the world's countries.

"Another cage supplier in Malaysia is Big Dutchman," Mr. Tan continued. Big Dutchman was founded in Holland, is today headquartered in Germany, and sells pig, chicken, and hen equipment in more than a hundred countries.

"Our slaughter equipment is from Holland," Mr. Tan went on. "Our egg grader equipment is Staalkat—also from Holland. Corn and soy for chickens is from America, Argentina, and Brazil. We buy from Cargill." Headquartered in Minnesota, Cargill is the largest privately held corporation in the United States, responsible for a quarter of all U.S. grain exports.

Asian factory farms, I realized as I listened to Mr. Tan, are essentially conglomerates of factory farms from Western nations. Every single thing is imported, from animal breeds to animal feed to animal cages. All the latest industrial technology is combined to confine farm animals to a degree not exceeded even in the West. As European consumers are growing skeptical and wary of factory farms, European equipment corporations are setting their sights abroad—to resounding success.

Mr. Tan and I stood at the mouth of the third aisle of cages. In the dust, dirt, and stink, my nose started dripping like a broken pipe. I began wiping it on the mask in my hand until the mask became as drenched as a washcloth. Mr. Tan, comfortable behind his mask, tried not to look disgusted.

At my insistence, he and I walked through an aisle of cages. The necks of most hens were bare of feathers from pecking and wire grating, and the gaping pink skin looked like an open wound. As we passed them, the hens squawked like they were possessed, their wings flapping like fans. They jerked back and forth in their cages like yo-yos, hurling feathers and feed in all directions. With every step I took, I felt like I was treading into the eye of a hurricane.

I shut my eyes against the dust, and I raised my hands to my ears to block out the manic, castigating cackles of the hens. I wished I'd worn my mask, and I wished I had goggles and earplugs to go along with it. I wondered what life was like for the hens, who breathed this poisonous air for their entire lives.

Mr. Tan was falling behind, unable to withstand the squawks and the dust. His eyes above his mask looked as terrified as those of a child trapped in a maze. He requested that we turn back. I agreed with relief, abandoning my self-assigned mission of traversing to the other end of the facility. Even walking from one end of a factory farm to the other is a challenge: the environment is so abominable that full bodily protection is required.

When Mr. Tan and I were again standing at the mouth of the third aisle, I asked him about something that was confusing me. "The hens were not reacting to the worker who was walking ahead of us in the aisle," I said. "Why were they reacting so strongly to us?"

"Because we wearing white shirts. Workers wearing yellow shirts. Chickens never seeing white shirt before. Not used to white shirts."

I was astounded. Mr. Tan's reply suggested that hens notice the color of worker clothes and react when there is a change in them. Hens display an emotional and cognitive response—fear and distrust—in the face of a novel situation such as strangers in new colors. As I stood there with Mr. Tan, I also found it ironic that hens, who have evolved to be so strongly perceptive to color, are given such little color in their lives—mostly just variants of gray.

Two rats scurried past Mr. Tan and me.

"We sure got rats here," Mr. Tan said. "They cause *Salmonella*."

Cage eggs are far more likely to carry *Salmonella* than cage-free eggs, according to many studies on the subject. The rats got Mr. Tan started on disease, a topic that visibly worried him. "Our mortality here is 13 to 14 percent for layers [egg-laying hens]. In some houses, it is 26 percent."

In other words, one out of every seven hens died in her cage—an extraordinarily high fatality rate.

"There are many diseases we have here in Huat Lai and in Malaysia," Mr. Tan continued. "One is IB."

Infectious bursal disease (IBD) is a highly contagious, deadly virus that proliferates in conditions of crowding and filth.

"One more is *E. coli. Lot* of *E. coli.*"

Escherichia coli (*E. coli*) is a large, dangerous group of bacteria with the power to debilitate and kill both people and animals. As with infectious bursal disease, *E. coli* occurs more commonly in unsanitary environments.

"And ND also," Mr. Tan lamented, shaking his head with despair.

He was referring to Newcastle disease, a highly contagious virus that is endemic in Asia, and whose mortality can be close to 100 percent in unvaccinated flocks.

"ND is big problem, number one problem. ND can cause mortality very high for both boilers [broiler chickens] and layers [egg-laying hens]. Whole region facing this issue—Indonesia and Thailand also. Very serious disease in this region. When chicken having ND, egg production is low and egg is bad. Chicken look sick. Chicken cannot drink. Chicken cannot eat. Chicken die."

"What causes ND?" I asked Mr. Tan.

"Stress."

"So, what do you do about stress?"

"We putting vitamins in chicken feed."

Vitamins? Was there no other way to reduce stress? How about getting the hens out of cages, since cages were their primary source of stress to begin with? Healthy, active hens have stronger immune systems and better disease resistance than cramped, caged hens. *All* the bacteria and viruses that Mr. Tan mentioned are scientifically known to be reduced by housing hens in clean environments and cage-free conditions.

"In 2004, we have H1N1 problem in this region." Mr. Tan continued, his eyes clouding with the pain of unhappy memories.

"For two months, the Singapore government banning the eggs from Malaysia. Everything stop. *Everything* stop. I crying every day then." He looked like he might cry again.

H1N1 was the same bird flu virus that had also infected the egg facility I'd seen in Indonesia. It is a testament to the strength of viruses and the weakness of agribusiness—with its immune-weak, industrially-bred, confined animals—that viruses are able to spread with such ferocity. They destroy not only animal and human lives—resulting in hundreds, thousands, and millions of deaths—but also business itself, resulting in the shuttering of borders and markets. But companies somehow still don't learn.

"Right now, Huat Lai have only layers [egg-laying hens] in cages," Mr. Tan said, brightening up. "But in three more months, all our boilers [broiler chickens] will be in cages also. Cages have good cost savings. With cages, we can keep more boilers in one house. In ten years, I think most boilers in Malaysia will be in cages."

In university, I had a friend who refused to eat fast food. "KFC and McDonald's have awful practices," she stated simply. She could not explain it to me further than that, and I had no idea what she meant. Until Malaysia.

When people crowd counters at KFC and McDonald's, when they sit down with their plastic trays at plastic tables, they do not simply consume a chicken drumstick or a hamburger. They consume an entire system.

Wherever fast food companies go, wherever they set up shop, they bring with them an industrial mindset and a mass-production methodology. They push animals to their biological limits. They concentrate ever-greater numbers of animals in ever-smaller spaces. They accelerate the transition from "open" to "closed" houses. They stock closed houses with industrial animal breeds and equipment from the west. The dark, filthy conditions under which animals are reared encourage the proliferation of myriad kinds of bacteria and viruses.

Fast food succeeds only on the basis of factory farms. KFC, for instance, is among the five largest integrators in Malaysia, and it is also the most popular fast food provider. Cobb chickens were invented by Tyson Foods in the United States, but in Malaysia, they are grown and slaughtered by KFC.

Fast food chains alter not just meat production but also meat consumption. People start eating more meat, and this greater amount of meat is produced more deplorably. The low prices at fast food cash registers are a direct result of callous indifference toward farm animals.

Factory farms are, in a sense, like palm plantations. "I think of palm plantations as good, bad, and ugly," the airport taxi driver had told me. Factory farms are good economically, bad ethically and environmentally, and ugly aesthetically.

A 1995 international survey found that 88 percent of participants could correctly identify McDonald's golden arches, compared to only 54 percent who could correctly identify the Christian cross. The survey's results suggested that fast food is the world's modern faith, colonizing and converting people everywhere, its gates in the shape of golden arches, its god not Christ, but Colonel Sanders. McDonald's has more than 35,000 locations worldwide today and KFC has more than 18,000. In many countries, there are more fast food restaurants than there are places of worship. In almost all countries, fast food restaurants are far more frequented than places of worship.

Fast food can be compared not just to faith, but also to countries and economies. McDonald's daily customer traffic of more than 70 million people is about 1 percent of the world population—greater than the population of the United Kingdom.

McDonald's $28 billion in annual revenues makes it the ninetieth largest economy in the world.

Factory-farm corporations everywhere—companies such as Huat Lai—are all more similar than different, modeled on Tyson in the United States. The seeds of factory farming and fast food have been scattered in the soils of the world's nations and are now propagating of their own volition.

But there's no reason for them to propagate. Nobody is forced to buy fast food.

"The first step toward meaningful change is by far the easiest: stop buying it," writes Eric Schlosser in *Fast Food Nation*. ". . . A good boycott, a refusal to buy, can speak much louder than words. Sometimes the most irresistible force is the most mundane."

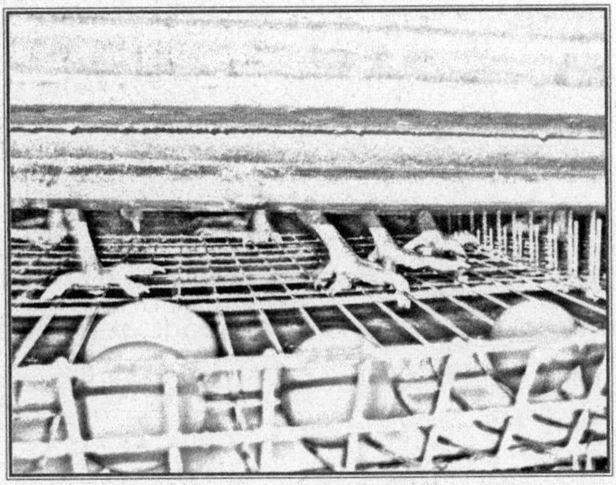

CHAPTER ELEVEN

LION CITY SINGAPORE

CRISPY PORK BELLY

The start of my stay in Singapore was not fortuitous.

I arrived late at night, exhausted, to find that all hotels were booked. Lugging my suitcase behind me, I went from one sliding-glass hotel door to another, only to be repeatedly told: "Sorry. We are fully booked."

Of the many hotels at which I inquired, just one had capacity, and for obvious reasons: it was located in the red light district. I got a small, shabby room and sank into a slumber for the night. I awoke the next morning to find that the air conditioner had dripped water onto my threadbare suitcase, further deteriorating my meager possessions.

My living experience in Singapore was more the exception than the norm. Located off the southern tip of the Malay Peninsula, Singapore is a clean, cosmopolitan country of slender skyscrapers. At one quarter the geographic size of Rhode Island, it is no more than a dot on the global map, and yet it is still a commercial and financial center—the New York City of Southeast Asia. Owing to its high level of economic growth, Singapore is described as an Asian Tiger (although the literal translation of its name means Lion City).

The country's motto, "Onward Singapore," reflects the national attitude. Singaporeans toil in high-rise offices, then shop in high-rise malls, then sleep in high-rise apartments. Life is orderly and regulated, and competitive and hectic. The country, however, is the perfect example of the maxim that money can't buy happiness. A worldwide survey in 2012 found that Singaporeans are the least happy people in the world—less happy even than the residents of war-torn nations like Iraq, Afghanistan, and Syria.

Like many industries in Singapore, agriculture today is outsourced. Most meat, milk, and eggs are imported from Malaysia and other countries.

As such, Singapore was unfortunately not a good place for me to study animal agriculture. I nonetheless made the best of it. There were no buses or taxis in the agricultural area of the country, so I hitchhiked my way to an egg operation. When I arrived, I was told that it had shut down years ago because of the same international H1N1 bird flu outbreak that had decimated the egg industries of both Indonesia and Malaysia. The incident highlighted, yet again, both the devastation and inevitability of disease at factory farms.

I found a second egg operation called Seng Choon. It was the largest farm site I'd ever been to, its property measuring fourteen hectares, an area equivalent to twenty-six American football fields. Even from outside, far from the entrance, its twenty long sheds stank of rotting eggs, rotting hens, rotting manure. Family-owned Seng Choon produced one eighth of Singapore's eggs by confining more than half a million egg-laying hens to cages, using "machines almost in the whole process" of egg production. Not only was the size and the "fully automated and computerized" system of egg production an imitation of the American egg industry, but the primary hen breed, Rhode Island Red, was imported from the United States as well.

Seeing Seng Choon, with its mass-production systems, I could understand why Singaporeans are second only to Americans in their low spending on food. Just one dollar of every fourteen consumer dollars in Singapore is allocated to food.

Deathly afraid of bird flu, Seng Choon's management treated me like I was a confirmed carrier of the virus, a walking pathogen. They asked me to leave immediately. I was not surprised at my lack of admittance, for just outside the gate stood a red placard with a drawing of an officer pointing a rifle at a man with his hands in the air. The cautionary picture was accompanied with the words: PROTECTED AREA: NO ADMITTANCE TO UNAUTHORIZED PERSONS.

Seng Choon's website expressed the hostility more mildly. "Before you enter, we must make sure you are also free of germs," it stated. "After all, you don't want to get those adorable chickens sick." Seng Choon's egg cartons were adorned with large labels of "Farm Fresh" and HIGH QUALITY & FRESHNESS. (I detested the word "fresh" by now, feeling allergic to it.) "When you are eating a Seng Choon egg," stated the website, "you are enjoying a pure product of nature. You are tasting freshness & quality that has been prepared for you with thought & love."

I wandered the streets outside Seng Choon for a while, then hitchhiked again. My driver, an Indian man in his thirties, treated me to a well-meaning lecture. "It's not safe for women to walk around alone in this area!" he said, looking agonized. "Recently, there was a report in the news that a woman was raped. . . ." He dropped me off at one of the few dairy farms left in the country.

The farm was uniquely illuminating, the polar opposite of Seng Choon.

It was an Indian Hindu farm, a fact evident from the outside, for a grand idol of a blue deity leaning on a cow graced the entrance. The cows at the farm were the cleanest I'd ever seen, their skin sparkling. Especially after Seng Choon, this was a culture shock. What was this?

"Hindus love and respect cows," I was told. "We are grateful to them because they give us milk."

Admirably, the farm *bathed* the cows every morning. They also refused to slaughter them, instead burying them when they died of old age. This was a whole different kind of relationship with farm animals, one I'd never seen before, one that involved doting instead of dominating. Even here, in Singapore, the land of industrialization, such a farm was possible—and thriving.

Frog with Dried Chili. Crispy Fried Live Frog. Signature Chili Crab. Deep Fried Prawn (Must Try!). Cereal Prawn (Must Try!). Curry Fish Head. Assam Fish Head. Crispy Fried Chicken Wing. Yum Yum Chicken. Duck Tongue and Jelly Fish. Peking Duck. Crispy Pork Belly. Sweet and Sour Pork. Hot Plate Beef Ribs. BBQ Roast Beef. BBQ Chili Squid. Hot Plate Deer Meat.

I browsed the menu in the restaurant, hoping desperately for something that did not contain animal flesh. I was left disappointed.

Though I'd had no trouble with food in Indonesia or Malaysia, many restaurants I visited in Singapore had not a single vegetarian option on their menu, not even a soup or a salad. I would have found more vegetarian options in American steakhouses than in standard Singaporean restaurants.

I was noticing a trend during my Southeast Asia trek. Wealthy people eat more meat. Meat consumption per capita is estimated to be 20 pounds per person in Indonesia, 115 pounds in Malaysia, and 160 pounds in Singapore. (A half-century ago, meat consumption levels were less than half present-day levels in each of these countries.) Indonesia's GDP per capita in U.S. dollars today is $3,500, Malaysia's is $11,500, and Singapore's is $52,000, about equal to that of the United States. Singaporeans are fifteen times as rich as Indonesians, in other words, and they eat eight times as much meat. As Dr. Shan of MARDI in Malaysia had expressed to me, "When people get rich, they put their money in meat."

In addition to Singapore's wealth, a second reason for its elevated animal consumption is its culture. Three fourths of Singapore's residents are Chinese, and most Chinese today eat lots of animals. The average person in China today eats 115 pounds of meat every year, about four times as much as only four decades ago. Animal consumption in China today constitutes more than a quarter of all global animal consumption and more than half that of all developing countries. China is today the world's largest animal eater by far. On a per-capita basis, Americans eat more meat than the Chinese, at over 175 pounds per person per year, but China's population is four times larger.

Trends in China occur faster than in the rest of the world, but they move in parallel. Worldwide meat consumption has increased by almost 3 percent every year in the last half-century. The lion's share of growth has come from developing countries, propelled by rising incomes and the perceived correlation between class and carnivorousness.

World trends of animal consumption are disheartening. A lone place of optimism is India, where meat consumption per capita has actually dropped in the last two decades. A half-century ago, the average person in China and India ate the same amount of meat. Today, the average person in China eats as much meat as about fifteen people in India. Indians have the lowest meat consumption in the world, at seven pounds per person per year. The United Nations Food and Agriculture Organization expects India to continue moving into the future as "a predominantly vegetarian society by international standards."

I made my way to Dubai after Singapore.

Dubai is the success story of the Middle East, a metropolis of soaring salaries and no taxes. It is home to some of the world's tallest skyscrapers, most expensive hotels, grandest malls, and most unusual property projects, including man-made islands. A land of expatriates, most of its residents speak English or Arabic. Whereas Singapore blossoms under a year-round tropical rainforest climate, Dubai blisters in the dry desert sun.

I went to the largest chicken factory-farm corporation in Dubai and spent hours talking to a white-haired, flush-faced South African executive, Mr. Booth, in his office. "If you want to enter one of our farms," he told me gravely, "you'll have to isolate yourself in a room for three days straight. You won't be able to leave the room for even a moment."

"I think I'd go mad if I did that. Can you make an exception for me?"

"I'm sorry, we can't. If you think that's bad, you should see the biosecurity for our workers. They're forbidden from leaving their accommodations for a full four weeks—the life period of the chickens. Our workers are isolated completely. Their food and water are delivered to them."

"Is such a high level of biosecurity necessary?"

"*Very* necessary."

Throughout my conversation with him, Mr. Booth spoke of workers as he did of farm animals. His company also treated them similarly, isolating them, forbidding them from leaving even for food. I realized then that not only does agribusiness ignore animal nature but also human nature. It's not normal for people to be on house arrest—or farm arrest—for four weeks. Workers will undoubtedly find ways to roam at night, when no one is watching. If the health of an entire system rests on utter isolation—of both animals and workers—it rests on the wrong foundation. If complete isolation *were* a possibility, bird flu outbreaks would not spread like forest wildfires.

"Everything we use comes here from all over the world," Mr. Booth continued. "We deal a lot with Europe. Our chicken breed is from Germany. Our feed mills are from Switzerland. Our slaughter equipment is from Denmark. Our cooling pads are from the States. The farms here are just like Tyson in the States. The way it works is that companies from abroad come here and do turnkey projects. They set the whole thing up. They build the chicken houses, they put in the feeders and bins. All we do is go there, turn the key, and stock the chickens."

In the Middle East, as in Asia, everything arrives from a different source, with the hope that it will all work together, like the parts of an engine. For a while, it seems true. The car that is agribusiness starts roaring with steam and rolling down the road at high speed. But when one stops and peers at the engine, one sees that it is greased by the wrong lubricants and exists on the verge of explosion.

CHAPTER TWELVE

THE MONSTERS OF MEXICO

A CONVERSATION WITH A VETERINARIAN

I'd learned a lot in Asia, but I hadn't had any epiphanies. I'd succeeded in collecting more pieces to my farm puzzle, but instead of becoming clearer, the picture had only become more complex and shapeless. Collection was useless without connection, I realized. After floundering around uncertainly, I decided to seek new soils, but a little closer to home this time. I went to Mexico.

The season was spring, and the day of my arrival was sunny. Mexico is a country of palm trees and beaches, and cities and resorts, and tacos and burritos. With a population of 120 million, it is the world's most populous Spanish-speaking country. Its economy is strongly linked to the United States. When it comes to agriculture, Mexico is perhaps historically important for having been the birthplace of corn, which today forms the primary feed for farm animals around the world.

As soon as I arrived at my grimy hotel, I went to the front desk and asked the man at the reception where I could find a farm. He'd never heard such a request before. It bewildered him. "Why aren't you going to the beach like other tourists?" he demanded indignantly.

He relented to pull out a yellow phone book and to look under the words *granjas* (farms) and *pollos* (chickens). Most places he called did not pick up the phone; those that did said, "We do not allow visitors." He tried a new strategy. He stopped cabs buzzing past the hotel outside and asked them if they knew of any farms. Three cab drivers said they did not. One said he did. But he did not; he took me to a petting zoo.

After more inquiries in more places over the next days, I finally managed to find my way into animal agriculture.

"Our company name is Keken," said Omar. "We are the biggest pork producer in Mexico—number one. This is one of our offices. As you can see, it is a business environment. We cannot give you much information. It is not normal for us to share information about us with others. But because you are a student, it is okay. We will tell you a little, but most things, we cannot tell you."

Omar was an executive at Keken in his mid-thirties, his title *Jefe Producción Exportación* translating to Chief of Export Production. He had fair skin, a round face, and long-lashed green eyes. He spoke English well because he'd lived in North Carolina for two years of high school. His employer, Keken, slaughtered a million pigs a year, a process that it described as "transforming them into meat products," achieved through "humanitarian sacrifice so that the animals feel no pain."

"We have many pigs," Omar went on. "A lot. We have at least a hundred farms and usually at least two thousand pigs on one farm. The market wants more ham—always more ham, it's never enough. . . . Keken is a total integrated company. We have our own production chains, our own farms, our own feedmill plant, our own genetics, our own slaughter, our own retail shops. We also have our own slaughter by-product rendering plant, which makes the raw material for other animal feed. We own everything. In Mexico, we sell ham in our own shops and also in supermarkets like Walmart and restaurants like TGI Friday's."

The largest American shopping and dining chains have almost as significant a presence in Mexico as they do in the United States. Mexico is home to more than 240 Walmart Supercenters, 150 Sam's Clubs, and 30 Costcos, as well as more than 500 Domino's, 380 McDonald's, 300 KFCs, and 180 Pizza Huts.

"We sell much more ham to Japan and South Korea than we sell in Mexico," Omar continued. "They are our big customers. We follow their needs closely."

The Americas are becoming Asia's pig factory farm, I realized during my conversation with Omar. Wealthy Asian countries—Japan, South Korea, and, as I saw firsthand, Singapore—are eating more and more animals, but they do not want to have factory farms on home turf, because they would then be forced to contend with the resulting air and water pollution from manure. They want to have their lunch and eat it, too, so they outsource their animal farms.

In addition to Keken in Mexico, the companies that I'd come to know in Canada—Conestoga Meat Packers and Maple Leaf Foods—also count Asia as a customer, especially the countries of Japan, South Korea, and China. Mexico has long been exporting pig meat to Japan and South Korea, and it started exporting millions of dollars worth to China in 2012 as well. Japan, South Korea, and China also form the largest markets for net imports of American pig meat today, and they continue to balloon over time.

The story of outsourced agriculture made newspaper headlines in 2013, when China's largest pig meat producer, Shuanghui International, announced its decision to acquire America's largest pig meat producer, Virginia-based Smithfield Foods. It was the largest takeover ever of an American company by a Chinese one. It was surprising, but also not surprising.

Shuanghui slaughters a startling 30 million pigs annually, in comparison to Smithfield's 28 million. Both Shuanghui and Smithfield are vertically integrated, controlling the entire process from breeding to feeding, slaughtering to rendering. They are also horizontally wide, owning myriad factory farms and slaughter plants, Smithfield operating not just in the United States, but also in Mexico, Romania, and Poland. In fact, it is Smithfield's Mexican factory farms that are thought to have been the birthplace of the deadly swine flu pandemic that swept the planet in 2009, killing up to half a million people in its wake.

For all practical purposes, the Shuanghui-Smithfield acquisition put it in writing: North America—all of it, from the arctic lands of Canada to the Midwestern soils of the United States to the dusty deserts of Mexico—is becoming Asia's pig factory farm.

The increased exports of pig meat are immensely profitable for American, Canadian, and Mexican corporations—they cannot keep the pigs coming fast enough—but they are awful for pigs, people, and the environment. More pigs are confined into ever smaller spaces to amplify production. "What's good for pork exporters may not be good for the United States," states an article in *Mother Jones* magazine. "More mass-produced pork also means more pollution to air and water from toxic manure, more dangerous and low-wage work, and more antibiotic-resistant pathogens. And that's just the beginning."

It really is. The fact is: North America is already suffering under the severe environmental strain of factory farms. A single factory farm can generate as much waste as an entire city. One adult pig produces about 13 pounds of manure a day. This means that 1,000 adult pigs excrete a total of 13,000 pounds a day. Over one year, the amount becomes five million pounds of excrement. This manure bristles with harmfully high levels of ammonia, hydrogen sulfide, nitrogen, antibiotics, and pathogens—which then end up in water bodies. According to the U.S. Environmental Protection Agency, factory farms are the most common culprit of polluted rivers, streams, and estuaries in the United States.

"Our goal at Keken is to get bigger," Omar continued. "We will keep increasing, increasing, increasing, every year. We want to stay number one in Mexico, and we want to compete with other countries in the world. We are big in Mexico, but Smithfield in U.S. is a monster compared with our company."

The word "monster" rang a bell. The man who'd told me about Keken—someone I'd met at an agricultural office—had said, "Keken is *too* big. It's a *monster,* a very big monster." And now Keken was calling larger producer Smithfield a "monster."

"No, you cannot visit one of our pig farms," Omar told me upon my asking. "You have to write a letter to the company. Even if *I* want to go to one of our farms, I have to write a letter. Special permission is needed for everyone."

I went next to a company called Crio, which had a headquarters that looked to be a bright yellow house. I showed the two watchmen at the gate a piece of paper that read "*Quiero hablar con alguien que hable inglés.*" I want to talk to someone who speaks English.

The watchmen waved me in to two receptionists. The receptionists waved me in to two employees. The employees waved me in to Dr. Julio.

Dr. Julio had a black handlebar moustache whose darkness was accentuated by his charcoal-gray hair. He was tall, with an oval face and a swift smile, and he was warm, pleasant, and attentive, the nicest and most informed man I met in Mexico. In both person and profession, Dr. Julio reminded me positively of Dr. Shan in Malaysia.

We got into his large pickup truck for the half-hour drive to Crio's operations.

"I was born in the north of the country," Dr. Julio said. "I became a veterinarian when I graduated from university in 1988. There are many veterinarians in my family. Two of my uncles are veterinarians; they are retired now but they were in dairy cows. Three of my cousins are also veterinarians. One cousin is in poultry like me. Another cousin is in pork and third cousin is in beef. I have two daughters; the big daughter is seventeen and she wants to be a veterinarian like me. She wants to be a veterinarian for cats and dogs because we have four dogs at home. Veterinarians for pets are doing well, phew!

"I learned English in night school when I was twelve years old. Now, my English helps me because I go to U.S. three or four times every year. I go to farm shows to buy farm equipment. I go to the big poultry show in Atlanta. You get good discounts if you go to the show. I ordered a lot of equipment for chickens when I went there. It's the same system here in Mexico and in U.S. Same farms, same machines, same process.

"'Crio' means child in Spanish. The owner of Crio wanted a name that is easy to remember. I have been at Crio for almost twenty years. Before that, I worked for Tyson for two and a half years after I finished university. Tyson is a *big, big* company. At Crio, we sell a lot to Walmart, Sam's Club, and Costco. We are fully integrated; we own full process. We make our own cages even. Today, I cannot take you to egg farms, but I can take you to broiler houses. We have about two hundred broiler houses, and a total of four million broiler chickens."

Paved roads gave way to a bumpy dirt path bordered by trees. At its end, we passed through a wide gate and arrived at a row of a dozen gray warehouses. They were the longest chicken sheds I'd yet seen. They were open from the two short sides, with a knee-high fence instead of a wall. We stepped over a fence and into a shed.

Standing amongst 30,000 chickens, my first observation was that the lines of red feeders snaking across the length of the place read "Chore-Time." Chore-Time was the same American company that had supplied the red feeders at Brick Roberts's turkey operation in Canada.

The chickens were sitting and some were clucking. "Chickens are very curious," Dr. Julio told me. "If you leave door open, they will be scared first, then some of them will go outside. They have never been outside. . . . I want to show you something."

Dr. Julio took me to the chicken shed next door. It looked the same from the outside, but from the inside, it felt and smelled very different, stinking more of ammonia.

"How much older do you think these chickens are compared with those in the other house?" Dr. Julio asked me.

"Maybe two to three weeks."

"No. *Four* days. These chickens are only *four* days older."

I knew this logically—I'd been told it repeatedly—but I still could not grasp it psychologically. It contradicted the laws of biology. It was like saying that a full-grown cat is only four days older than a kitten—it's a concept impossible to reconcile with reality. *Poultry Science Journal* has calculated that if humans grew at the same rate as present-day chickens, a human would weigh more than 600 pounds in eight weeks. Consumers often assume that the freakish growth of chickens and turkeys is a result of hormone injections, but it's not—it's their Frankensteinian genetics.

"When they are born, chickens are forty grams in body weight," Dr. Julio told me. "But in fifty days, they reach more than three thousand grams of weight. They are very, very heavy. But they have problems because of that. Big problem is heart size. Another is lung size. Chickens today are bigger than before, but their heart and lungs are same as before. This means there is not enough blood, not enough oxygen to the heart. Chickens suffer cardiac arrest."

As we stood there, Dr. Julio asked me "Can you see the dust?" He waved his hand though the faint brown glaze, dispersing its particles, which danced around us then settled in the air in suspended stillness.

Dr. Julio looked down at the floor; I followed his gaze. Manure had solidified into pebbles, packed down by all the tens of thousands of feet that trod it. He picked up a pebble of excrement and crushed it between his fingers into a drizzle of crumbs. "We sell this chicken shit," Dr. Julio said. "Some of it is used to fertilize fields, and some of it is used as feed for cows."

Chicken excrement fed to cows? Feeding one animal the *excrement* of another? That was a whole new level of perversion, and I found it unsettling that Dr. Julio, an otherwise intelligent man, could speak so casually about such a practice.

Cows are fed chicken waste not only in Mexico but also in the United States. Because poultry litter costs less than corn, the American cattle industry feeds as much as a million pounds of poultry litter to cattle each year. This is distasteful no matter how one looks at it, but especially so when set beside the fact that chickens are themselves fed cattle remains in the form of rendered meat and bone meal. Chickens excrete the cow parts that they eat, and these parts are fed back to cows in the form of chicken feces. The strange cannibalism chain creates an alarming danger of disease, including mad cow disease, for which reason Canada has banned the feeding of chicken litter to cows.

A former leader of National Farmers Union-Ontario has spoken out against the practice, arguing: "There are of course many concerns associated with eating beef fed on chicken 'crap.' The more we learn about disease transfer—such as influenza viruses, the more we realize how at risk we are in our modern world. Chicken feed has as one of its ingredients cattle in the form of crude protein from meat and bone meal. . . . American cattle are potentially eating feed that could potentially be a link to BSE [mad cow disease]. . . . [We should] expose this practice for the dangerous and foolish risk it is."

Just at that moment, Dr. Julio and I both noticed a chicken. He was in the strangest, most grotesque position in which I've ever seen an animal. His legs dangling lifelessly above him, he was on his back, his head somehow folded *underneath* his back.

Dr. Julio strode up to him and picked him up by his wings. "I will have to cull," he announced dryly. He turned away, his back to me, and his arm jerked in the motion of wrenching—breaking the neck. He dropped the chicken to the floor.

The chicken thrashed about. He flapped his wings. He moved his legs. He tried to stand. He swung his neck. He sank into a slow death. Two chickens, waddling by, settled down on either side of him.

"Chicken industry has changed a lot, and it will change more," Dr. Julio said. "Biggest change will be cages. Many people say now that in ten to twenty years, all broiler chickens will be in cages, like all egg-laying hens. But I don't like cages. It's more stress for animals."

It was a breath of fresh air to hear Dr. Julio acknowledge that cages *do* cause stress. This fact was as obvious as the hair on my head, but nobody I'd met in agribusiness so far had actually *said* it.

"Most of our egg farms have four levels of cages," Dr. Julio went on. "But now we have made two new farms with *eight* levels of cages. The new cages are from Chore-Time—I made the deal with them a year and a half ago at the poultry show in Atlanta. The cages are very, very expensive, but we will recover the investment in only three years."

Dr. Julio showed me photos of the cages on his phone. The cages were shiny, metallic-gray, spanking-new skyscrapers that promised to take factory farming to a whole new level.

It grew dark outside as the sun set. Yellow lightbulbs hanging above us switched on automatically, to ensure that the chickens would continue eating and growing through the night.

As Dr. Julio drove us back to the city, he got a phone call from his boss, the owner of Crio. When he hung up, he told me "The owner wants us to increase our broilers by sixty percent in the next four years. That is a fifteen percent increase every year! It is very, very difficult."

"Will it happen?"

"Yes!"

CHAPTER THIRTEEN

BAREFOOT IN BEAUTIFUL BELIZE

LIVING WITH MENNONITE MISSIONARIES

It thunderstormed on the two ferries I took out of Mexico to Belize. Water fell like cold fire, pounding the ferries into submission.

The reward for the rage was a rainbow that appeared, as if by magic, to form a graceful arch of pink, teal, and orange above the water, a handshake of friendship between the ocean and the elements above it. The sun peered out from behind the clouds, and turquoise waters shimmered like crystals under its newly scathing gaze. In the depths of the waters swam golden, green, and indigo schools of fish—forming their own underwater rainbows—alongside sharks and stingrays that drifted through the currents like ghosts. White-winged seagulls navigated through the winds like paper planes.

Belize is a haven for the over-worked and the under-sunned, a restful and leisurely *ashram*, except that the God here is not Buddha but Bob Marley. Belize has a resident population comparable to that of Honolulu, Hawaii, and its culture is similarly constructed on sandy beaches and the American vacationers who seek them. Belize was previously called British Honduras, and the English queen continues to reign as monarch into the present today, playing a mostly ceremonial role.

I'd thought my ferry rides from Mexico to Belize were turbulent—subject to the fury of clouds that opened and closed like trapdoors—but they were like a spa in comparison to the bus rides of Belize. Bus rides were like *Orange Is the New Black*—a dog-eat-dog bubble of prisoners trapped together. Passengers *ran* onto buses, pushing and shoving, making a mad dash as though they were escaping a fire. I couldn't do it. I missed the bus.

The second bus rolled in, and I would have missed it again had a sturdily built woman in the crowd not taken pity on me and grabbed my wrist, thrusting forth through the throng with me and my suitcase in tow. She pulled me onto the bus, but there were no seats left. I resigned myself to standing tiredly for two and a half hours, finding myself reminiscing about something I'd thought I'd never reminisce about: the air-conditioned buses of Mexico, where I'd sat next to snoring men and the TV screens had played awful action movies like *The Fast and The Furious*, made worse by their dubbed Spanish.

Two women on the bus permitted me to perch on the edge of their seat. The bus wound through neighborhoods of pastel-painted houses and orange orchards and palm trees. Ranges of mountains rose in the distance like hazy apparitions, their foliage a menagerie of all shades of green: dark, bright, lime, olive. The national motto of Belize is appropriate to its landscape: "Under the shade I flourish."

As I looked out the bus window, I noticed something conspicuously absent in Belize, something favored in Mexico but shunned here, something that formed also a crucial difference between Indonesia and Malaysia: fast food. There were no gaudy M-shaped arches in Belize, no posters of Colonel Sanders's face.

"The only food chain here from America was Subway," said the woman sitting next to me. "A Subway restaurant opened; there was interest in it for two months—only because it was new—and then people stopped going. It closed down. Belizeans don't like fast food."

Belize seemed like the right place for me.

When I disembarked from the bus, I took a taxi to the Mennonite farm where I would be staying. But the farm gate was locked, and no one was home. Five dogs appeared behind the bars of the gate and barked at me uproariously. I plopped down on my suitcase and waited.

"I'm sorry, dear. I hope you haven't been waiting at the gate too long and the dogs haven't scared you too much. We were at church. It's Sunday. Our church is in the mountains, along a dirt road, and it takes us time to get there and back. But we never miss it. We haven't missed it once in years. Come in. Let's sit down and have a cup of tea. I'm Geraldine Lemon."

Geraldine Lemon had sun-reddened cheeks and dark brown hair streaked with tendrils of gray. Her hair was rolled back into a matronly bun clamped with a black bonnet. She wore a lilac ankle-length dress with a pattern of faded white flowers. Her smile was soft, but her voice was harried at the edges and her face cooled at the blue eyes, which stood as unchanging in temperature as a stretch of deep water.

"The right temperature of the home is maintained by warm hearts, not hot heads," informed a placard in the living room. The house was furnished with a long wooden bookshelf, flowing curtains adorned with parrots, and rows of colorful greeting cards hanging on a line with clothing pins.

"Tell me about you, dear," said Geraldine Lemon. "Do you have children, by the grace of God? Are you married? By what age do you envision yourself married, dear? What kind of man will you marry? Oh, you're not sure, are you? Well, will you marry someone who *believes*? Believes in *what*, you're asking?! Dear me, in *God*, of course. In the *Lord*. The *Almighty*. I have to ask: do you yourself believe in the Almighty? Do you pray?"

I made a movement of my head, a sort of wiggle, that I hoped could be interpreted as either affirmative or negative. Geraldine Lemon's wise eyes saw it for what it was.

"When I was younger, I was a bit of a . . . let's say, for lack of a better word, a *hippie* like you. I thought I knew everything. My parents were Christian, but I rebelled against everything they taught me. The sixties were a time of rebellion, dear. I was born in a suburb of England, but in my rebellion, I moved to London and later to Paris. I did yoga. I became a vegetarian. I met my husband somewhere along the way. We traveled across the U.S. in a van for a year and a half.

"When we got tired of seeing waterfalls, we decided to come down to Belize and settle down here. We didn't like the powdered milk we were getting in Belize, so we sold our van for a dairy cow. We met Mennonite missionaries here, and they were good to us, so we converted: we became Mennonites.

"We, Mennonites, put all our faith in the Almighty. We believe that the kingdom of God is not the world's kingdom, so we don't participate in worldly systems. We don't pursue education, we don't go to war, we don't vote. We're often farmers precisely in order to separate from the world. Mennonites do most of the farming in Belize.

"There are many kinds of Mennonites, dear. Some use more technology and others use less. Our house, as you see, is moderate. We have most technologies except a microwave and a television—that's where we draw the line. The important distinction between different kinds of Mennonites isn't what they have or don't have, but how conservative they are. There are more conservative Mennonites and there are less conservative Mennonites."

I was hoping Geraldine Lemon would say she fell under the classification of less conservative Mennonites.

"We don't like the less conservative Mennonites, dear," she pronounced.

I was wearing a navy-blue, knee-length dress with a scarlet belt at the waist, and I had latched onto my head a huge, navy-blue head bow that I thought lent a whimsical hint of old ways. I was dressed like this specifically because I thought it would appeal to Mennonites. It didn't.

"We, Mennonites, wear *ankle*-length dresses," Geraldine Lemon told me pointedly. "And we never, ever take the Lord's name in vain. We never say something like, 'Oh my God.'"

I'd already said it at least once; it formed one of my signature expressions, unfortunately. I didn't know how I'd avoid the term—and what the consequences would be if I didn't.

"I've lived in Belize for over thirty years now, for more than half my life," Geraldine Lemon continued. "My husband passed away eight years ago. I have eight children—five daughters and three sons. Most of them were born in Belize, and they grew up on our dairy farm. They're eighteen to thirty-seven years old now, and most of them are married. My three sons work together as Realtors in an office at the bottom of the hill. I live at this dairy farm with two of my daughters: my youngest daughter, Abbey, and one of my middle daughters, Nancy. One of my granddaughters, Katie—the youngest daughter of my eldest daughter—lives with us, too. It's the four of us you'll be seeing a lot of—myself, Abbey, Nancy, and Katie. Three generations of Lemon women. Oh, here they come."

Geraldine Lemon's zeal for God was like a raging fire, and I leapt up and raced outside when I heard her daughters and granddaughter.

Eighteen-year-old Abbey was slim and lovely. Her voice was surprisingly burly and husky—a distinctively crisp sound bursting with passion and optimism—a voice in which an entire personality was contained. Twenty-six-year-old Nancy was great in girth, her face flat, her shoulders slack, her dragging feet suggesting a weariness with life and its mundane challenges. Ten-year-old Katie was part Mennonite and part Mayan—part white and part brown—and she was beautiful, with tanned, rosy skin, and a charming, shy temperament.

Like their mother, Abbey and Nancy wore floral cotton dresses and black bonnets. Each woman had her own accent. Geraldine's was fully English, Abbey's was English-influenced, and Nancy's was Caribbean—"yaa, man!" Granddaughter Katie spoke too softly for her accent to be accurately identified.

"Nancy's in charge of house chores like cooking, baking, and cleaning," Abbey chirped to me. "I'm in charge of farming. I'll show you the farm and you can meet the cows! We have a herd of thirty cows and calves. We have twenty-one pastures, and we move the cows from one pasture to another every day so the grass has time to grow."

The pastures were resplendent, circled with low hills and sprinkled with majestic trees of palm, coconut, orange, mango, guava, and banana. Several black, white, and brown cows stood in the shade of the trees, their tails swinging in the breeze like whips of freshly washed hair. Other cows lay in pairs, leaning against one another, looking, as Abbey described, "lazy and languid in the afternoon."

Tall, starch-white birds with long golden beaks and golden manes stood in the grasses, peering into the cows' faces, appearing to gauge their next move. "Those are cattle egrets and they want to eat the cows' ticks," Abbey explained to me. "It's a symbiotic relationship they have. Cows want their ticks eaten and cattle egrets want to eat them."

"I grew these cows up," Abbey continued merrily. "I washed them every day. I braided their tails. When my favorite cow died, I cried for days. Those two cows there, Aida and Anita, are best friends. They're always together. Belgian and Brady are friends, too. I see them hanging out more these days. Cows like hanging out with cows their own age, not younger or older cows."

Abbey herded the cows into a new pasture, bounding with a swift, sprightly gait through the tall grasses. The cows followed her easily. Those who were to be milked followed her into the milking parlor, an outdoor area with stalls. One cow was slow, lingering to chew grasses. *"COME, Inkie!!"* Abbey yelled at her in encouragement. Recognizing her name, Inkie ambled over faster, coming in to stand next to Stephanie.

Abbey had prepared a treat for the cows. Piles of moist orange peels, leftovers from the orange juice she'd squeezed for breakfast, lay in the feed troughs. Abbey had drizzled the orange peels with a rich syrup of molasses and sprinkled them with a yellow powder of wheat bran and cottonseed. The treat was, in human terms, equivalent to a pile of pancakes drizzled with chocolate syrup and sprinkled with sugar. The cows relished it, devouring the peels in gulps. "Some piles of orange peels are bigger than others," Abbey told me, "because some cows like oranges more than others."

I was astounded at Abbey's knowledge of her cows. She had not only given them names, but she knew their friendships just as well as her own, and she even knew their relative desire for orange peels.

Ducks and cormorants swam in a hyacinth-covered oval pond at the farm, and other birds flew low overhead. "At our farm, we have cattle egrets, pigeons, blackbirds," Abbey enumerated, "great-tailed grackles, kiss kiddies, northern jacanas, vultures, laughing falcons, hawks, storks, woody ground doves, white barn owls. . . ."

With all the birds dropping by, the Lemon farm was not just a *farm,* but also a bird-watcher's paradise, a sanctuary not just for domestic animals but also for wildlife. It was a delicate dance of balance, an ecosystem in which the well-being of each was intertwined with the well-being of all. I'd never seen such a biodiverse farm before; I was amazed.

Abbey and I were traversing her mother's blooming flower garden to return to the house, when I heard something. A low chorus, a clucking twitter. . . . Hens?

Auburn-feathered Rhode Island Red hens lived in two rustic enclosures in flocks of twenty-five. The enclosures were the size of small rooms and were sunlit and fresh-aired. They were adorned with nest boxes, furnished with a wide wood ladder whose rungs served as perches, and carpeted with a plush layer of sawdust. Hens walked and pecked, scratching out shallow burrows in the sawdust and settling into them.

Abbey went home and returned with wet, leafy folds of lettuce from the kitchen. She was trailed by little Katie, who held a woven peasant basket. Abbey gave the hens the lettuce—they ate it as eagerly as the cows had eaten orange peels—while Katie and I collected eggs from nest boxes by hand, placing them gently in the basket. When we were done, Katie put the basket down and did what she'd *really* come to do: lavish the hens with affection.

She picked them up one by one and caressed them as one strokes a cat. They were calm and unresisting to her; they knew her. Their feathers were soft and lush to the touch. So different in their temperament and appearance were these gentle creatures from their caged counterparts I'd seen in other countries, that they could just as well be a separate species.

In Canada, cage-egg producer Brick Roberts and his best friend Paul had said, "Hens are happy in their cages with all their friends." What the Lemon hens did next refuted their claims definitively.

"We open their doors at five every evening," Abbey told me. The hens streamed out as soon as she opened the door, forming a red-brown wave of feathers, spilling into the garden like children into a playground. They started pecking for worms in the grasses, scratching up dirt, and strolling about. It was hen heaven.

The Lemon sort of hen housing system is called an aviary, and it suits both the physical and social needs of hens. Importantly, it can be imitated on a large scale with ease. Instead of two enclosures of hens, there can be twenty-five or fifty. Instead of twenty-five hens per flock, there can be fifty. In total, an aviary operation can have thousands of hens (though probably not tens of thousands), and they can walk and roam instead of festering in a cage.

The Lemon dairy and egg farm was humane, environmental, beautiful, and eye-opening. It was exactly what I'd been seeking all over the world. I could hardly believe I'd finally found it.

Each of my days with the Lemons began with a breakfast of pineapple and papaya drizzled with lemon or orange juice.

Katie and I would then accompany Abbey as she milked cows or mixed dirt—all the while humming and singing, bouncing and skipping, as if nothing in life could be more enjoyable than cows and dirt. I would ask her about the colorful birds flying overhead and Katie would eat guavas off trees.

After morning chores would come a long, jovial walk along a dirt road bordered by shrubs and sprawling trees. I would walk in flip-flops; the Lemon women would walk barefoot. "Mennonites like being barefoot," Geraldine told me. I tried it, too, but the gravel and pebbles were hot and harsh against the soles of my feet, and I returned to my footwear promptly.

In the afternoon, we would have lunch, often a salad prepared by Nancy. The Lemon women ate meat only rarely. Almost all their meals centered around fruits and vegetables, to the extent that they did not even notice I am vegetarian; we were all eating the same foods. My favorite part of the day was late afternoon: tea time.

Over a pot of tea and a plate of delicious, fudgy brownies, Geraldine would read aloud to us from a novel, favoring Jane Austen, Charles Dickens, and James Herriot. "I don't like reading modern books, with their bad language and themes, and with their four-letter words and their 'Oh my God!'" she explained to me.

After tea time, missionaries would descend upon the house in droves. Abbey introduced me to a missionary couple with three children in their arms. "Sonia's studying animal agriculture," she said.

"Not *agriculture*, Abbey," her mother Geraldine corrected sharply. "Sonia's studying animal *husbandry*."

I'd met well over a hundred people in agriculture by now, and no one but Geraldine Lemon had made that distinction. But "animal husbandry" *is* an essential distinction. It entails a whole different kind of relationship with farm animals, one that implies a joined destiny and a view toward stewardship and shepherding instead of domination and exploitation. Animals are not meat, milk, and egg machines to be pushed to the limit but are, rather, comrades in the circle of life.

For their opinions on animals, however, among other matters, Mennonites are ridiculed. They are viewed as dinosaurs, as relics of the past persisting, who knows why, into the present. They are people who have not been left behind, but who have chosen to stay behind. For this slap to the face of society, they cannot be forgiven.

I'd been among the ranks of such close-minded critics when I'd arrived, but I was no longer. I found it courageous and marvelous that Mennonites farmed as they did, eschewing the agricultural rat race toward industrialization, favoring husbandry over machinery. In their opinion, and in my newly formed opinion, the past and the present need not be at odds. The present can become stronger as far as it draws from the best of the past.

"If we open a quarrel between past and present," said Winston Churchill, "we shall find that we have lost the future."

"I'm just a silly woman," I heard Geraldine Lemon say more than once, discounting her own opinions when speaking with male farmers or missionaries.

"I want to get married so I can have the guidance of a husband," young Abbey told me.

In the Mennonite culture, women are sometimes viewed (and view themselves) as lesser than men. But interestingly, it was only in the Mennonite culture that I came across a women's farm: the Lemon farm was managed by three generations of women. This state was not for want of sons, for there were three sons in the family. It was because women *wanted* to work with animals.

Women are a potential solution for agriculture, I realized.

One study finds that men and women overlap by only 10 percent in their personality trait distributions, the greatest difference between them being sensitivity. Another study published in the journal *Brain and Cognition* demonstrates that men and women differ in their capacity for compassion, defined as "a moral emotion related to the perception of suffering in others, and resulting in a motivation to alleviate the afflicted party." The study shows that men and women differ not only in how they process compassion psychologically—but also *neurologically*.

Other studies also conclude that women are more empathic than men, both psychologically and neurologically. Men and women differ so sharply in their personal and emotional characteristics that some researchers describe them as possessing "two human natures." Differences between men and women are thought to arise from divergent evolutionary sexual selection pressures in the areas of mating and parenting.

Whether or not society is aware of the scientific evidence on the differences between men and women, it overwhelmingly agrees with it. A 2008 Pew research poll found that 80 percent of Americans believe women to be more compassionate than men.

Differences between men and women extend definitively to their opinions on farm animals. Surveys from around the world show that women are more concerned about farm animals than men. Women are also more likely to favor better treatment for them and to support increased protective legislation.

Gender differences translate to eating habits as well. Women of all countries eat less meat and less fast food than men, and are more likely to be vegetarian than men. In addition, research shows that women are more likely to purchase organic food, to think about food safety, and to evaluate health, nutrition, and sustainability in making their dining decisions.

Some say that the gender bias toward men in agriculture is dictated by the physical nature of the work, as in industries like construction or manufacturing. Though this was once true, it is today less so, the Lemon farm being a case in point. "Farming's not like it used to be where you needed a strong back and a weak mind," a farmer is quoted as saying in *The Progressive Farmer.* "Now it's better to have a strong mind and weak back."

If women farm workers are rare, women senior executives are an endangered species. I came across not one over the course of all my investigations. My experience, as it turns out, is reflective of the industry as a whole. As of 2014, Cal-Maine Foods, the largest egg producer in the United States, counts just one woman on its twenty-member leadership team. Chicken producer Tyson Foods has one woman on its fifteen-member executive team. Pig producer Smithfield Foods has one woman on its seventeen-member executive team. At each of these three multi-billion-dollar factory-farm corporations, women constitute one out of every fifteen to twenty senior executives. Women have been unable to crack through the "grass ceiling," as some in the industry put it.

And yet, although most food is produced by men, most of it is purchased by women. This is the case in several other industries, too—men produce, women purchase—but it is more pronounced when it comes to animal raising (men) and grocery shopping (women). The product packaging, with its bucolic images, is designed more to attract women, though. There are few women in agriculture, but agriculture markets largely to women—to great success.

In the wake of the Wall Street crisis of 2008 and 2009, of which I was myself a casualty, articles appeared in the media with titles such as: "Does Wall Street Need an Estrogen Injection?" "Testosterone and high finance do not mix: so bring on the women," and "Mistresses of the Universe." In the latter article, Nicholas Kristof of *The New York Times* wrote: "Banks around the world desperately want bailouts of billions of dollars, but they also have another need they're unaware of: women, women, and women. . . . Aside from issues of fairness, there's evidence that the result is second-rate decision-making. . . . The number of studies reaching similar conclusions from different directions is striking."

The different directions include science, retail, government, law, and police work, among others.

Michael Lewis, author of *The Big Short: Inside the Doomsday Machine,* a book about the economic crisis, expressed in a 2009 *Vanity Fair* article: "One of the distinctive traits about Iceland's [economic] disaster, and Wall Street's, is how little women had to do with it."

His top recommendation for preventing future financial catastrophes is to fill 50 percent of risk positions in banks with women. "I've always thought that Wall Street would be much less insane to the extent that it let women in," Michael Lewis said at a talk. He continued that the number and the role of women in finance can "very easily change," and that, "when it does, we'll all be a little safer."

Food will be safer, and animals will live better, if more women work in agriculture.

"We sell our milk and the yogurt we make from it at our farm," Geraldine Lemon told me. "People buy it from us directly. But everything else—our eggs, Nancy's brownies and breads, Abbey's granola—we sell at the farmers' market every Saturday."

The retail chain makes a difference to the production chain, I realized during my stay with the Lemons. Vendors at farmers' markets tend to be small-scale local farmers who are often humane and sustainable, like the Lemons. Shoppers can converse with farmers directly and ask them questions, without a separation of hundreds or thousands of miles. Farmers' markets encourage connection and care on the part of both sellers and buyers. "You support a lot of values when you shop at the farmers' market," states food industry writer Michael Pollan in an interview.

Consisting of anywhere from a handful to hundreds of stalls and stands, farmers' markets today are found in all kinds of locales, whether cities, suburbs, or villages. There exist tens of thousands of them around the world. Large supermarkets have a "high volume, low margin" model, as egg producer Brick Roberts had described to me, focusing on quantity over quality. Farmers' markets have fundamentally the opposite model—low volume, high margin—which means that they sell fewer foods but of higher, more careful quality.

In France, *marchés* are so common and colorful on cobbled city and village streets that I was driven to write a paper about them during my French-language study-abroad program. In the United States, the tradition of farmers' markets was brought over in the early days from Europe, and it spread with settlement. The number of farmers' markets in the U.S. has jumped almost five-fold in the two decades from 1994 to 2014, rising to more than 8,000. Hundreds of new farmers' markets sprout up across the country every year.

The dramatic resurgence of farmers' markets reflects a positive shift in public attitudes. Consumers are starting to care more about the treatment of animals and the sustainability of land. They are beginning to view grocery shopping as an exercise in living their values. They are realizing intuitively that their values are better served at independent, original farmers' markets than at brightly lit, big-box megaliths.

Before my stay with the Lemons, I'd never considered that a shopper's place of purchase can make a difference. I found this uplifting because it means that even if we do not want to change what we eat—if we do not want to be vegetarian, for instance—we can simply change where we shop and what we buy to start making a difference. We are not just what we eat, but also where we buy what we eat.

I asked the Lemons to point me toward their friends' animal farms. "We'll all come," they said instead.

There was a buzz of excitement as Nancy, Abbey, and Katie got ready, treating the day as a school field trip. They wore fresh dresses and made new braids and buns, adorning them with flowers from the garden. Katie inserted a mesmerizingly sweet-smelling white frangipani into my hair as well. Nancy drove the truck, I sat next to her in the passenger seat, and Geraldine, Abbey, and Katie sat in the back. Geraldine sang about God's love, her voice rising and falling with passion. "He will provide . . ." "I'm *free*, free from this world . . ."

We passed several herds of cows and sheep on hilly grasslands, their dark hues creating an attractive design on the carpet of green. Cows in Belize rarely number more than twenty to a herd, and they graze on an average of one to two acres each. But it's not just cows—the entire agriculture industry of Belize is small-scale. The total number of all the chickens, turkeys, and pigs in all the farms of Belize is equal to that generally found at a *single* operation in the United States, Canada, and Mexico. Belize shows that it is possible to feed an entire nation, albeit a small nation, while heeding the values of animal compassion and land conversation.

Not all is beautiful in Belize, however.

We arrived at an egg farm belonging to a friend of Geraldine's. A Mennonite woman of immense proportions, Laura resembled a *Shrek*-like ogress. She and her husband produced eggs as contract growers for a poultry company in Belize called, inaccurately, Quality Poultry Products. Their egg farm was elevated three feet off the ground, and under its wire floor were accumulated four rows of excrement that looked like ranges of hills in their peaks and valleys. Seeing the abundance and shape of manure, I knew by now exactly what I would see inside. The Lemon women didn't.

Laura's hens lived in four rows of wire cages. They were Rhode Island Reds, like the Lemons', but they were thin and weak, with pimple-pink, exposed skin and brittle, broken feathers. There were about a thousand of them. A passionate debate flared at their sight.

"Poor chickens!" exclaimed little Katie softly. She looked ready to cry, clutching her grandmother Geraldine's hand for support.

"This is *cruel*," Geraldine berated her friend Laura. "Our hens run outside in the evenings. They walk about. I feel so sorry for your hens."

"It doesn't matter how we keep hens," Laura countered with a laugh. "They don't have brains like us." She raised her banana-wide finger to her temple. "They can't *think*."

Geraldine would spend much of the next day imitating Laura—raising her finger to her temple and cackling, "They can't think, they can't think"—concluding always with a despairing "How *awful*! How *cruel*!"

Nancy began explaining the cruelty of cages to two of Laura's teenage daughters, who'd accompanied us in. "Hens need to move, they need to walk, that's why they have legs."

Abbey's reaction was strongest of all. "I am in *shock*," she said, her hands clasped to her heart, her expression agonized. "I'd heard about hen cages before, but I hadn't known they're *this* bad. I just can't believe it. These cages are tiny—they're nothing."

This was the first time I'd entered a battery-cage hen facility with humane farmers. Their prompt, powerful reaction indicated precisely how revolting it was.

Laura was not offended by her friends' opinions, but nor was she influenced by their concerns. Talking to her was like lecturing a brick wall; criticisms bounced off like deflated squash balls.

CHAPTER FOURTEEN

AG-GAG AMERICA

GLITTERING GREEN MOUNTAINS

In recent years, several American states have passed or tried to pass laws to ban reporters and investigators from factory farms. *The Guardian* in the U.K. has described these ethically and constitutionally questionable laws as follows:

"Multiple states have passed what are known as 'ag gag laws,' designed to penalise investigative reporters who explore conditions on industrial agriculture operations. . . . These laws are a significant threat to the freedom of the press, and it's rather remarkable that they are being allowed to stand. More than that, they threaten the health and safety of consumers, in addition to making it difficult and sometimes impossible for consumers to make educated choices about the sources of their food. The US should be in an uproar about ag gag laws, and it's not. That's a telling reflection of attitudes about agriculture . . . Attempts to raise awareness about the issue are often met with indifference; they are not as interesting and seductive as celebrity scandals, evidently, even though they are far more scandalous, and impact people's lives more immediately and directly."

The Atlantic has expressed: "When it comes to the meat supply . . . America appears none too troubled by the prospect of its blindfolding; the nation would rather take its chances with *E. coli* than risk channel-surfing into a slaughterhouse." The editorial board of *The New York Times* has written: "Instead of ag-gag laws, we need laws that impose basic standards on farm conditions and guarantee our right to know how our food is being produced." "As you next cut into a steak or crack an egg," advises the editorial board of *The Washington Post*, "ask yourself why an industry that claims it has nothing to hide demands protections afforded to no other."

I'd skipped the U.S. on purpose on my way south from Canada to Mexico and Belize. The reason was that I was terrified at the idea of investigating agriculture in the U.S. I feared that I would end up in jail for having done nothing at all. And yet a part of me realized that no analysis of animal agriculture could be complete without a dive into the burly, bubbling belly of agribusiness, the kingpin of farm animal confinement, the United States. Factory farms are just as much a conceptual and literal export of the U.S. as action-packed Hollywood, lyrical pop music, and chocolate-chip cookies.

I decided to be smart about things; there was no need to be a hero. States that have ag-gag laws on their books are Idaho, Iowa, Kansas, Missouri, Montana, North Dakota, and Utah. I decided to give these seven states a wide berth. The states I would investigate, I decided, would be Vermont and California.

California was an easy choice because, in addition to being the home of actors and entrepreneurs, the state is ranked second in the nation for animal agriculture, preceded only by Texas. As for Vermont, it is the primary agricultural state in New England, and it suited me geographically because I was planning on attending my five-year alumni reunion across the state border at Dartmouth in New Hampshire.

Dairy forms the primary farm-animal sector in both California and Vermont. California is America's largest teat, supplying one fifth of the country's milk. I expected that the two states—one among the largest in the country and the other among the smallest—would provide me with an overview of American animal agriculture from coast to coast, with an emphasis on dairy.

But things started going downhill before I even landed in Vermont. The airplane ran low on oxygen. My ears began to explode with pressure, and swallowing felt like pulling a trigger on a gun. Yellow masks dropped down with pops and I put one on.

Eventually, we landed, still breathing, still living, and I was now forced to face a new fear. A car rental. For a quarter of an hour, I sat in my rented car, a white Ford Focus, unable to move, my hands leaving damp fingerprints on the steering wheel. As I turned the key in the ignition, I was consumed with nightmarish visions of my driving instructor, who'd panicked more with each of the twenty-five long driving lessons I'd taken with him. "The car is out of *CONTROL!!*" he would yell hysterically. In my last lesson, he'd announced, in conclusion, "You're not improving with the lessons. You'll never learn to drive." I fired him in a fury.

I did eventually take a driving test and, despite multiple errors (like my inability to park), I did somehow pass it. My instructor's final forecast, however—"You'll never learn to drive"—kept playing in my mind like a bad song. Driving, to me, seemed almost as challenging as sprouting wings and soaring off into the sky like a bird.

With its glittering green mountains, curving roads, and scarlet cottages, Vermont was breath-taking, among the most beautiful places I'd ever seen. It was endowed with a special sort of splendor, one that made me stop my car in the middle of the road and stare, enamored.

I managed to drive to the dairy bed-and-breakfast where I was staying, belonging to a husband and wife and their thirty-year-old son. This was their story, as the wife, Liz Derek, who had a glamorous look with bob-cut blond hair and luminous blue eyes, told me:

"My mother used to make and sell burgers. I didn't want to help out, but I had to. One day, Bob, now my husband, came for a burger at my mother's stand. That's how we met. Our first date was on a cattle truck! I was nineteen, he was twenty-five. We got married three years later and had two kids, a son and a daughter. Our son works here. We're the fifth generation; he's the sixth. The farm's been in my husband's family since the mid-1800s. Our son had better get busy making the next generation!"

Bob Derek, an enormous man with a flushed face and freckled arms, found it funny that I'd attended Dartmouth. He found the concept of university funny, actually. "We've never been to university," he said. "We've never been abroad either. We've never been interested in school or travel."

The bedrooms in the house were designed with floral bedspreads and billowing curtains, and the living-room walls were decorated with black-and-white photos of stern, thin-lipped grandparents. In an ancient, abandoned barn on the property hung two charming black-and-white signs: SAVE THE FAMILY FARM and IN THE LAND OF MILK & HONEY.

This land of milk and honey contained 70 milking cows who formed part of a total herd of 140. At any dairy farm, the "milking herd" consists of only those cows giving milk at the time, and the "total herd" consists of all cows and calves, and tends to be double the size of the milking herd. Though "milking herd" is a more commonly used number in the dairy industry, "total herd" is more relevant in getting a sense of the sum of animals.

The dairy cows were black-and-white Holsteins. Holsteins constitute more than nine tenths of dairy cattle in the United States and Canada because of the exorbitant quantity of milk they produce. The Dereks' cows lived in a red barn connected to hilly grasslands with a view over the valley. The cows strolled over the hills, and birds flitted by over their heads, singing as they went, coming to rest on beech, maple, oak, cedar, and chestnut trees.

"Our cows go out whenever they like," Bob Derek told me. "We turn them out to a new pasture every day so the grass has time to grow back. We have eighteen pastures. We manage the pastures, and the pastures manage the animals."

This farm closely resembled the Mennonite Lemon farm I'd stayed at in Belize. The dairy herd was small, and it was shifted to a new pasture every day. The farm was thoughtful and sustainable.

"On our grazing and grass setup," Bob Derek continued, "cows get exercise, they eat well, and they're healthy. At large American dairies, there are too many cows in too little space. Farmers get away with it by feeding them antibiotics. . . . Let me explain the difference to you in your terms. Feedlot farmers think of a cow as a stock; they want a quick return on their investment. We look at a cow as a bond instead of a stock. We're not pushing the cow hard on a day-to-day basis, and so she lasts longer, and we get more milk long-term. It's win-win for everyone."

The bond-versus-stock analogy was illuminating, forming the difference between the long-term and short-term perspective. What was a large dairy like? I wondered. I'd never seen one before, and drove out to find one.

Cows were being milked by machines managed by an old man who had wiry hair and walked with a limp. The milk frothed white and creamy into large round glass cases whose painted lines marked its weight, as the bars of a thermometer measure temperature. As the old man turned around in the performance of his functions, he jumped to see me standing behind him.

I asked him for a tour of the dairy. He dispatched his son, Nelson, on a walkie-talkie. Nelson had gray stubble and gray eyes, and he was trusting, patient, and pleasant, talking to me like he already knew me. I liked him instantly.

Nelson's title was Assistant Herdsman, but it was outdated, for the cows lived in feedlots, and he had no herding to do. Much like the Dereks with whom I was staying, Nelson and his father had owned a dairy farm of seventy milking cows. *Un*like the Dereks, they'd sold their farm and started working at this large dairy six years ago. This was a family farm as well, purchased eighty years ago and now in the hands of the third generation. But it had 480 milking cows, summing to a total herd of close to a thousand. It was immense.

The cows lived in long red sheds, under shadowy roofs, and they met my eyes with their drowsy, long-lashed ones. Their udders brimmed with milk, lined with purple and green branch-like veins. Half of the cows lay in their feedlot stalls and the other half limped and doddered unsteadily from stall to corn. They stuck their heads out through metal bars to eat the rolling piles of ground, brown corn on the other side.

As I looked at them, I realized that dairy cows have been turned into egg-laying hens. Their diet is corn, and they eat it by sliding their head between bars. They formed a dull, depressing sight.

Because the cows got no exercise in their feedlots, they were weak and lame. About a quarter of feedlot dairy cows in the U.S. are estimated to be lame. Also, because of their poor health, American dairy cows are considered "spent" at only four to five years of age, a fraction of their natural lifespan of twenty or so years. They are trucked to slaughter to become low-grade hamburger.

The term "feedlot," I realized, is exactly appropriate because it is a "lot" where cows are "fed," rather than a land where they nourish themselves, and where the land is in turn replenished by their grazing. Feedlots originated in the 1950s and '60s as a way to raise cattle cheaply on corn, and they resulted in the shifting of cows off grass and onto concrete. The term "feedlot" is used more often with regard to the beef industry, but it has become increasingly relevant to dairy.

There was something grotesque and garish about the cows: half of them had tails that had been mutilated to stumps—six inches long rather than the usual length of three feet. They looked like severed limbs.

"We were docking all the cows' tails until recently," Nelson told me, "but animal rights people started making a big deal about it. I don't think it's a big deal. It's a personal preference whether to dock tails or not. A lot of dairy farmers do it. But in California, they outlawed tail-docking in 2009. We figured they might outlaw it in Vermont, too, at some point, so we've stopped doing it in advance. That's why half the cows have tails and half don't. We haven't docked the tails of the younger ones."

Tail-docking is a painful practice. A rubber ring is tied near the top of the tail to cut off blood circulation to the lower two thirds, which atrophies and detaches, or is cut off. Tails are docked to make milking more convenient: cows no longer swish their tails in the faces of milkers, because they no longer have tails. But the aftermath of tail-docking is almost as distressing as the days-long process of atrophy itself. Because cows no longer have tails, they are powerless to fan flies away, and become lathered with them.

Anywhere from half to four fifths of American dairies dock the tails of some or all of their cows. In Europe, in contrast, tail-docking is widely considered barbaric and is prohibited by law in Denmark, Germany, the United Kingdom, Scotland, and Sweden.

Dairy cows nudged one another out of the way to sniff and lick Nelson and me as we leaned against the railings. Their stalls seemed bedded with something dark brown. It was neither straw nor sawdust. It looked like . . . mud. Or . . . manure?

"We bed our stalls with dried manure," Nelson confirmed. "Cow bedding has traditionally been straw or sawdust, but a lot of large dairies today are starting to use manure. We put thousands of gallons of manure into a digester every day, then we spread it out in cow stalls after it's digested. The manure solids save us a lot of money on bedding, and the manure gases make us a lot of money by producing electricity. And we get subsidies for our digester. There's an awful lot of grant money in this; the government's been aggressive about promoting digesters."

Savings on cow bedding accrue to dairy-farm owners, in other words, but costs of manure digestion are distributed over taxpayers, cows, and consumers. Digesters cost one to two million dollars per farm, which means, startlingly, over $500 per cow. Cows lay in their own dried, recycled waste day and night, resting their heads on it, sniffing it perpetually with their powerful noses. A cow's persistent contact with bacteria can also lead to a greater chance of contracting mastitis and other udder infections, which can affect the safety of milk.

Digesters are promoted to the public on the basis of reducing odor and pollution. But though digested manure is less likely to pollute groundwater and rivers, it pollutes cow stalls. Also, when it comes to manure digestion, as to other areas of animal agriculture, the system of government subsidies is geared toward benefiting large industrial operations.

"Large anaerobic digesters are used to make factory farms more viable," states Energy Justice Network. The Sierra Club also opposes subsidies for digesters, explaining: "We would rather see the farmers who are profiting pay for them. If any other waste producer was producing a steady waste stream, they would not be rewarded."

I heard a machine-like sound and looked up. A teenager on a lawnmower whizzed by on the grass outside the feedlot. This was ironic; shouldn't cows be the ones trimming the grass, by eating it? Why were they being fed corn and confined to feedlots when there was abundant grass to eat?

There was a second sight outside even more bizarre. A red-haired woman in sunglasses—Nelson's wife—walked by briskly, making rounds of the feedlots with a little boy in tow and a toddler in a stroller. Feedlots in rural America are considered so *normal*, so ordinary, such a routine part of life, that children are not taken for a stroll of a park, but for a stroll around the feedlots.

There was a further tragedy in the mundane scene. Nelson was a dairy farmer, his father was a dairy farmer, his grandfather was a dairy farmer, and, most likely, his son would also grow up to be a dairy farmer. But since he was a baby, all that Nelson's son had ever seen was hundreds of cows confined to concrete feedlots, bedded on manure, their tails docked, poking their heads out through metal bars to eat corn. Wouldn't Nelson's son believe that cows *belong* in concrete lots? Wouldn't he believe that they have no instincts and interests, that they're not animals but automatons, machines that eat corn through their mouths and produce milk through their udders?

Calves at Nelson's dairy farm lived in a separate, small, ramshackle barn. They were crowded, their legs smeared with one another's excrement. Strangely, their faces and backs were scarred with coarse, gray, crater-like crusts. The crusts resembled the dry, rough scars that form after burns, before the skin heals. Cringing but curious, I reached out a hand to touch a scarred calf's face.

"This is a disease called ringworm and it can be transferred to humans!" Nelson exclaimed. My hand snapped back. "Ringworm's not a worm," he continued casually. "It's a fungus that eats the skin and hair of calves and cows. It's a skin condition. It's really contagious; calves and cows get it from one another. It takes months to heal. Sunlight is the best cure for it, because funguses and molds don't like sunlight—they thrive in the dark."

Nelson's tone was nonchalant, but I was shocked. Such is the grazing, outdoor nature of cows and calves, such is their need for sunlight, that when denied their nature, they actually get sick. It's not just their muscles and bones that suffer in feedlots but also their skin and hair. Cows are more susceptible to ringworm when they are deficient in certain essential vitamins—vitamin D, from the sun, and vitamins A and E, from grass—just as people are susceptible to scurvy when they are deficient in vitamin C. The existence of a fungal disease such as ringworm highlights that cows have not a preference, but a biological *need* for sunlight and grass. But dairy farms still crowd them into shadowed lots. The outcome is that ringworm explodes to become a contagion.

Nelson and I walked through another feedlot and came across a pathetic sight. Three calves—black and white, meager and thin, two of them newborn twins, still wet from their birth—lay together in a heap like a pile of refuse. As I squatted down to look at them, they attempted to wobble up onto long, trembling legs, and collapsed.

"These calves are male," Nelson said. "They were born today. We're sending them to slaughter as soon as possible, whenever the next truck is, in the next day or two."

The calves hadn't even learned to walk yet, but a knife would soon slit their throats. They had spent nine months in their mother's womb—the same length as a human pregnancy—but they would spend no time at all on earth. These were the "bob" veal calves that I'd learned about earlier, at the veal farm in Canada.

Nelson and I walked into a smaller area, where we came across a sight almost as horrible as the heap of calves. A cow lay in her stall, blood dripping from her nose, her eyes wide open but unseeing. "She was put down," Nelson whispered to me. "We shot her with a gun in the back of the head. She was sick. We don't know why she was sick—it might have had something to do with mastitis." Nelson realized it was irrational to whisper about a dead cow, as if she could understand him, and he cleared his throat to loudly declare "We'll compost her later."

Many cows at Nelson's farm were infected with mastitis. They were distinguishable from the others by means of red bands around their ankles. Their milk was unsafe for human consumption. Clinical mastitis is the most commonly reported health problem in the American dairy industry, affecting one out of six cows. Manure and unsanitary conditions make mastitis more likely.

Nelson asked me my purpose. "I'm studying international animal agriculture," I told him without missing a beat. "Trends and practices."

"So you're an . . . agricultural economist?" he asked.

This was not far from the truth, I realized, surprised that I hadn't recognized it before. As a university student, I'd studied economics and government because these had seemed to me the areas with strongest influence on world affairs. My experiences at farms were confirming this supposition, for economics and government are the strongest drivers of agribusiness.

It is the economic desire to cut costs and make profits that determines animal living conditions. It is politicians who choose to side with agribusiness, subsidizing factory farms and feedlots and passing ag-gag laws instead of regulations to protect farm animals, consumers, and the environment. Change cannot come without addressing economic and policy drivers any more than a fruit can blossom without a root.

"Our kind of farm is traditional here in the U.S.," Nelson informed me on my way out.

Animal agriculture has changed so much in the last decades that even its language has changed. Factory farms are so entrenched today that they're described as "traditional" even though there is nothing traditional about them. Grass-based farms, in contrast, are described as the "new way," even though they're the old way, the true way.

Marketing, no matter, remains traditional. This dairy farm's milk was sold in white bottles with cartoon pictures of plump cows grazing on deep-green pastures.

I also investigated the American egg industry. Rather, I *tried* to investigate it. I got no further than the offices of two of the last remaining egg producers in New England. The first of them was managed by a pleasant, charming man in his mid-thirties who had an admirably incessant smile.

"We produce organic eggs. You must have passed by our egg barns when you came into the office. In addition to producing eggs on site, we also contract with twenty-six farms in the area.

We have 170,000 hens on site and we have 400,000 hens with our contractors. The way it works is that we supply our contractors with hens and we tell them how to build their barns. All our eggs are brown, because people associate the color brown with better health, like whole-wheat bread compared to white.

"Fifteen years ago, my parents decided to switch from the traditional hen cages to organic farming. We made the switch because we were getting squeezed out by our competitors, by the bigger farms in the industry. When you have half a million hens, you can't compete on price with a farm that has seven million hens. The only way to survive is to turn to an alternative market, to distinguish yourself not on price but quality.

"Every year, the number of farms in all sectors of animal agriculture drops, as farms consolidate and get bigger. There are only a hundred and eighty egg companies in the entire country. Your survival depends on whether you're able to cut costs by cents. Things are getting worse in the egg industry over time.

"The one and only good thing that's come out of industrial animal agriculture is that farms today can have more women than before." I was pleased that he felt the same way I did—that more women would be beneficial for the industry. "The U.S. is the perfect example of free-for-all agriculture, of agriculture with no controls whatsoever. Free-for-all agriculture is bad for everyone. It's bad for farmers, it's bad for animal welfare, it's bad for the environment, and it's also bad for consumers."

I returned to my car sobered up by the conversation, aware anew of the intensity of agricultural competition. On the way, I passed the long red hen warehouses outside the office, each of them bordered on one side by a strip of grass. I'd been refused permission to enter the farm buildings because of "biosecurity," but as I stood beside my car and gazed at them, I contemplated the question: What does Big Organic mean for egg-laying hens?

The answer: it means something, but not as much as one would hope.

Of the 170,000 hens on site, no more than 70 hens—less than 0.05 percent—were outdoors, availing themselves of the agreeable afternoon air. There were three reasons for this. First, in order to reach the grass, the hens had to climb out one small door, then *another* small door—which was not even parallel to the first small door. Second, the strips of grass were meant by their narrow width to accommodate a total of no more than a couple thousand hens, a paltry percentage of the vast sum. Finally, hens that are not permitted to step out when they are young usually do not step out when they are older, because they are not accustomed to the outdoors. It is partly with this knowledge that many organic producers give hens an outdoor option only when they are adults and less inclined to use it.

In sum, every effort was made to ensure that hens did not use the limited outdoor range they had. In contrast, the small flock of hens I'd seen at the Mennonite Lemon farm in Belize had enjoyed stepping outdoors. This was because they'd been foraging since they were chicks, they had plenty of outdoor range, and, to reach the gardens, they had only to walk out one straightforward, human-sized door.

Cage-free eggs are always, *without exception*, better than eggs from cages, but *how* much better is an important question. The answer: it depends on the intention and attention to detail of the operation.

Driving on Vermont's rural roads, surrounded by green mountains, I stopped my car occasionally to ask people if they could point me to a dairy farm. "Go to Dubois," they said. But despite the certainty in their voices, the look in their eyes spelled warning.

Dubois Brothers began with thirteen dairy cows close to a century ago. Today, three generations later, it has a milking herd of 1,250 cows and a total herd of 2,500. Like Nelson's dairy, multi-million-dollar Dubois is a happy recipient of government subsidies toward the digestion of manure and its use as cow bedding.

A Middlebury College student wrote a report about Dubois in 2010. "Feed is delivered into the stalls via tractor," he described. ". . . Milking is done around the clock. Immigrant labor is employed. . . . Dubois farm bears little resemblance to the dairy structures that people associate with Vermont."

Dubois consisted of several endlessly long feedlots with metal roofs and high green walls draped with a tarp to prevent cows from looking outside and strangers from looking inside. From within the bowels of its feedlots echoed the clanging noises of metal banging on metal. Goosebumps rose on my skin, for workers would surely order me to leave. This place was far larger and darker than Nelson's dairy. I walked in nonetheless.

Cows were packed together in groups of hundreds like schools of fish, each unable to move without bumping into others. The sun, high and bright in the sky, illuminated the continent but not the feedlots, for their black tarp obstructed its rays and cast the cows in oblong shadows. Vermont, with its glistening mountains and rolling grasses, was tailor-made for cows, but the cows could not benefit from the wondrous greenery all around them. They could not even see it.

Cows stood in slippery liquid piles of waste, and they lay in it as well, smeared with it like bread with butter. Each of their tails had been sliced into a stump. Many cows, especially younger ones, were infected with ringworm and had sick, scaly faces. The acidic odor of ammonia from manure shot straight to the back of my throat.

It was cows, not men, who were the source of the metallic clanging, I saw with relief. As cows leaned through metal bars to dip their heads into piles of corn, the bars shifted and banged against one another. Pigeons perched overhead on high, wooden beams. They, urban dwellers, had found the slums of rural areas—feedlots—and settled there definitively. There were so many that there was no avoiding them; one defecated onto my head.

Young feedlot cows lumbered over to me and strained forward to lick my hands and hair. The difference between younger and older cows was evident not only in the size of their bodies and udders, but also in their gait. Older cows walked carefully and quietly, like old women, some of them limping. They'd lost their footing too many times on manure-laden concrete, and they'd grown scared and scarred with age.

Two cows fell in front of me. The first was young and rose immediately. The second did not. She collapsed in a way that I had not known it was possible for a cow to collapse. Her fall was not a slip as much as a split—a splat—her hind legs splaying outward at a right angle like a frog's. As I gasped, she turned around on the ground instead of getting up. Twisting her body slowly, she started crawling on all fours, sliding like a snake.

All the cows in her pen now stared at her, still and silent, none eating, none drinking. Two of them approached her and nuzzled her face. She kept crawling. I realized she was crawling in a direction away from me—my presence was creating stress. I walked down the feedlot and returned several minutes later. My departure turned out to be a mistake, however. She'd crawled into a stall, and I could no longer tell her apart from all the others in stalls.

Shaken, I walked through the other feedlots. Each of them was identical and indistinguishable from its neighbor, except for one. It was a closed place whose windows were boarded up with inflated white plastic. I assumed that it was a corn-storage unit.

After I'd been traipsing about Dubois for half an hour, I unfortunately came across someone. His skin was brown and he had a sharp long nose accentuated with prominent black eyebrows.

"I Oscar," he said in a heavy accent.

"Where are you from?" I asked him.

"*Méjico.*"

Oscar was thirty-two years old. He'd lived in Maine for two years, and in Vermont for two years, and he would be returning to Mexico in a year.

"Why are you going back to Mexico?" I asked him.

He struggled to find the words, and shrugged when he was unable to. "Sorry, no English. . . . You talk *español*?"

"Sorry, I don't." He looked crushed. "But my brother speaks Spanish!" I said in an effort to console him. His face beamed like a lightbulb.

Oscar didn't mind my presence at Dubois. "You is very nice," he said. "You want see babies?"

I nodded, though I didn't know what he was talking about. He pointed to the boarded-up building that I'd assumed to be a corn-storage unit.

Not only was the place boarded up, but it was barricaded by a crimson gate that had to be *lifted* to enter, requiring considerable manual strength. Without Oscar, I would have been unable to proceed. The gate was there to prohibit entry, and the windows were stuffed with inflated white plastic to prevent even the merest of peeks. Dubois really did not want anyone looking in there, let alone stepping in. This fact made me both very curious and very nervous.

When Oscar opened the next door, my eyes took a moment to adjust to the dark.

The place consisted of a dozen black rectangles, out of which emerged low *moos*. Approaching, I saw that each rectangle was a row of ten calves in tiny, individual wire cages. Each row of cages was wrapped in black tarp, narrowing the calves' existence to a rectangle. Two square flaps had been cut out of the tarp for each calf to eat and drink from a black pail and a red pail that hung in front of his cage. The place seemed to be a white veal facility, or at least constructed in imitation of one. I sought clarification, but Oscar didn't understand what I was saying.

There were 120 calves in total. They were small, curled-up bundles of black and white with big ears and big eyes. Some of them wobbled up when I extended a hand. The littlest calves stuck their heads out through the flaps of their cages and sucked on my pants. Other calves were lethargic and unreactive, appearing to lack the energy and the will to move. They raised their heads slowly to glance at me, then set their heads down with finality.

Many calves had a deep, raspy cough that was eerily human in sound. Others suffered from a slimy, mustard-yellow diarrhea. Most calves were coated with flies, which buzzed around them like bees around a beehive, but the calves were indifferent to them, and also to the rats that scurried among them.

Oscar led me to a storeroom to the side. Its door was scribbled on with blue and black marker; it served as an instruction board for Oscar and other Dubois workers. Numbers 1 to 12 were listed, each referring to one of the twelve rows of calves. Next to four of these rows was scribbled the word *AGUA*. The word means "water" in Spanish, and was written on the door as such because Dubois's Mexican workers spoke so little English that even the word "water" was unknown to them.

AGUA was written next to only four of twelve rows; this was because only one-third of the calves were to receive an adequate amount of water. Water is viewed scientifically as an "absolute necessity" for calves, as for other animals, and its scarcity is associated with calf scouring and stress, but Dubois calves, like many other calves in the industry, were deprived of water so that they would eat more corn and gain weight faster.

On the door, the letter "D," for diarrhea, was scrawled next to some of the rows as well. Medications were also listed: Tylosin, for pneumonia, and Lincomycin, an antibiotic invented for pigs but used at Dubois for calves, despite medical directions to the contrary. There were also other drugs listed whose scribbled names I could not decipher. The door, a window into an unreal realm, said more still: calf 6986 was blind, and calves 7378, 6875, 6926, 7124, and 7161 were dead.

The calves who were alive at Dubois survived on the basis of extensive medications and inappropriate antibiotics.

As I walked back to my car, waving good-bye to Oscar, I realized that dairy, practically the only sector of animal agriculture that still employs human beings at all, is almost barren of them now. At even such a gigantic farm as Dubois, I came across no one other than Oscar.

Dairy cows, animals who have until recently stood outside resolutely, their heads bowed to the grass, their tails swinging in the breeze, have today been defeated and denied. They have no grass and they have no tail. Only one out of ten American dairy cows today grazes outdoors at all. The last sector of pastoral agriculture has collapsed, with no one to witness its silent demise.

I visited a second New England egg farm, where I was greeted with notably less enthusiasm than at the first. The place was family-owned, and I spoke to the wife, who wore eyeglasses and knee-length shorts.

"I only talk to people who make scheduled appointments," she told me. "I won't be able to spare more than fifteen minutes for you. . . . No, you can't walk through our egg barns; we never permit that. To tell you our story in short: we're a second-generation farm. My father-in-law started the egg farm with two hundred hens in the 1940s, after the Second World War. My husband and I took over in the eighties, and my children will take over when they get older.

"The egg industry of Vermont is tiny today. When we first got into the business, there were seventeen egg producers in the state. Now we're the only commercial producer left in the state. Until recently, we were competing with farms in the Midwest and in Maine that had two to fifteen million hens each, and that were sending their eggs here, to Vermont. We couldn't compete with their prices. It just wasn't possible. We needed a survival strategy, and we came up with one. We decided to focus on being a local Vermont product, on marketing ourselves not on low prices, but on being 'local.' We survived right by the skin of our teeth. People are becoming more interested in 'local' over time.

"We have sixty-five thousand egg-laying hens in total. Most of them are in cages, but twelve thousand are cage-free, kept in a storage barn. Europe's banned battery cages for hens, but I don't think that's logical. I don't think it's going to happen here in the U.S., certainly not in my lifetime."

The marketing of this farm, like most industrial farms, was a sham. Company vans read CAGE FREE, even though it was less than one out of six hens who were cage-free. "What comes first, the chicken or the egg?" asked the website. "Our hens come first," it answered. "Our hens are treated with extra special care." At the hen facilities, however, exhaust fans circulated steamy, stinky layers of dust, and mountains of droppings lay accumulated at the back, littered with dead hens and rats, buzzing with black layers of flies.

Of the two egg producers I visited in New England, one was "organic" and the other "local," but they were not so because of any personal passion or philosophical bent. Their decisions were survival mechanisms, effective forms of market differentiation. They began espousing the values of "organic" and "local" only after they were able to profit off them, not before. There was something sad in this; it was like discovering that a favorite singer's heartfelt songs are not heartfelt at all—that the singer doesn't even believe in love, but croons about it only because it sells albums.

The word "local," in particular, is inane when it comes to industrial agriculture. The term says nothing about animal or environmental ethics; it says only that a factory farm, with its tens of thousands of confined animals and mountains of manure, has installed itself in the neighborhood.

I drove past and walked through many more Vermont dairies, gleaning several trends in the industry, and also myself.

First of all, my driving was clumsy, my street smarts minimal. I steered anxiously with sweaty palms. I was often lost. I was occasionally going the wrong way on a one-way street. I was treated to impatient honks and dirty looks and sometimes middle fingers raised in the air. I could not fathom my GPS. I thought 13:00 meant that my arrival time was thirteen minutes, but it actually meant one P.M., which meant that I would expect a drive to be thirteen minutes long, but would end up finding that it was an hour or two in length.

I would enter and exit dairy farms with only cows as my witness. At those farms where there were human beings, it was often the case that all workers were Mexicans, and most didn't speak a word of English, making conversation impossible. They would greet me with *"Hola!"* and I would reply with the only Spanish I knew—*"Hola! Como está?"*—and that would be the end of it. More than two thirds of farm workers in the United States are Mexican, which means that Spanish is just as necessary in American agriculture as it is in Mexico.

At some dairy farms with white workers or owners, I was interrogated like an established convict, being asked questions that were odd attempts at entrapment. "You're not with any funny group like PETA, are you? You're saying no, but that means you've *heard* of PETA. So what you're saying is, you've *heard* of PETA, but you're not *with* PETA?"

Another lesson I learned (again) is that the term "family farm" is interchangeable with "factory farm," and thus says nothing about animal welfare or environmental stewardship. Ninety-nine percent of American dairies are said to be family-owned. Also, farm descriptions such as "second-generation" and "fifth-generation" are meaningless. Whether farming is a family tradition says nothing about whether the family farms in a traditional way.

"When we moved here thirty-three years ago," one dairy farmer told me, "there were eleven dairy farms here in our valley of six towns. We're the last dairy farm left now. Dairy farms are closing and the remaining ones are getting bigger. Wherever you go, that's what you'll see."

That's what I did see. FARM FOR SALE signs littered Vermont's countryside. It was inevitably small farms that were for sale and it was inevitably to large farms that they were sold. Vermont dairy farms have been closing their gates every year since World War II. The total number of dairy farms in the state has plummeted from more than 11,000 in 1950 to fewer than 1,000 today.

The same trend is visible throughout the United States. A dairy farm is like a movie rental store: you don't know if it's going to be around next year. In the two decades from 1992 to 2012, the number of American dairies dropped by more than 60 percent. Three quarters of the dairies that closed their gates sold their herds to larger operations. As such, the number of cows per farm continues to increase every year; American dairy herds have grown by an average of 142 percent from 1992 to 2012.

Farm acquisitions are not just a conquest but a conversion. As a small sect integrates into a larger religion, small farms convert and integrate into larger operations. "Get big or get out" has been the industry slogan since the 1970s, coined by U.S. Secretary of Agriculture Earl Butz, who had a soft spot for industrial animal production. Several American farmers repeated this slogan to me dryly.

Occasionally at dairy farms, I would see a few cows outdoors on a paltry patch of grass or dirt, and I would ask why they had been permitted a small spot to roam when the majority of their brethren were enclosed indoors. Then, even the most hardened men would acknowledge, with a sheepish expression, "Pasture is good for their feet and legs. It's easier on them to be on grass than cement."

Studies show that cows on pasture have better gait and strength than cows on concrete, and that pasture can even cure lameness and help cows recover from hoof and leg injuries. Cows on pasture are healthier than feedlot cows, and their milk is also healthier. The grass they eat imbues their milk with vitamins and omega-3 fatty acids.

Usually, before coming upon a dairy, I would *know* I would come upon a dairy. I would know this not by telepathy, nor the stench of manure, nor the sight of cows. I would know it by the sight of corn. Wherever there are animal farms in America, there is corn. Its glimmering green stalks ascend to the sky, seven feet in height, rippling in unison like waves in the winds, yellow cobs of kernels sprouting at the peaks.

Corn and its derivatives form the basis for processed foods like candy, Coca-Cola, and cake. For cows also, corn is the bovine equivalent of candy, inferior to grass nutritionally. It is also inferior to grass physically, for cows can graze in fields of grass, but not in fields of corn. In its shape and structure, the corn crop is simply not conducive to grazing. As such, grass gives cows not just health but also a home; corn stalks, in comparison, could just as well be shackles around cow legs.

Corn has changed, "and much for the worse," writes Michael Pollan in *The Omnivore's Dilemma*, "the lives of several billion food animals, animals that would not be living on factory farms if not for the ocean of corn on which these animal cities float."

Large dairy farms tend to be more alike than not, I noticed in Vermont. They tend to have more than one location, and each location tends to have several feedlots. Inevitably, feedlots contain too many cows, living in too much manure, eating too much corn, with tails that are too short and ear identification numbers that are too high.

Dairy is reflective of the entire animal agriculture industry and is alarmingly among the better sectors of it. In the last few decades, dairy has transitioned from pasture to feedlot, but for pigs, chickens, egg-laying hens, and turkeys, the transition was completed long ago. Dairy cows today are semi-enclosed, but other farm animals live trapped in factories without even windows. The average pig farm in the United States has more than 5,000 pigs. The average American egg facility confines more than 300,000 hens to cramped, stacked cages.

Dairy cows are the most fortunate farm animals in the U.S.—the most "high-end," due to their exorbitant milk production—but even their lives are replete with suffering.

In the dairy industry, keeping herds apart is considered essential in keeping disease at bay. By all evidence, however, it fails. The majority of dairy farms I visited were severely afflicted with ringworm, to the extent that that they'd created large, separate pens for infected cows and calves. And yet the method of ringworm contagion was obvious even to the uninformed. Like chicken pox, the crusty ringworm calluses are itchy, and they make infected cows and calves want to rub perpetually against one another and the bars of their pen. The constant rubbing enables calluses to spread from neighbor to neighbor, from one pen to another, from one herd to another, and especially from older to younger, the latter with weaker immune systems.

I feared ringworm would pass also to me. The slightest twitch in my skin or scalp would initiate a foreboding in me. In the dairies of Vermont, I understood for the first time why airport entry forms in the United States and Canada ask passengers to check off whether they have been to farms. The U.S. Customs Declaration form asks you to reply with a "yes" or "no" to "I am (We are) bringing soil or have been on a farm/ranch/pasture" and also "I have (We have) been in close proximity of (such as touching or handling) livestock." The fine for a failure to declare a visit to a farm can be up to $1,000 for a first offense, and potentially as high as $50,000 for further offenses.

The American customs form asks nothing of contact with firearms or nuclear or biological weapons. One is to infer consequently that animal farms pose the greater danger today. The question of contact with livestock, which most passengers check off without a thought, is not frivolous but is, rather, a vivid indication of the extent of disease inflicting the industry.

I noticed that the number of wild birds flitting through dairy farms is inversely proportional to the number of cows on the farm. The reason is that there is nothing for birds to do on lands built solely on corn and concrete—there are no weeds or worms for them to seek and peck. On large dairies today, gone are the symphonies of sparrows, the songs of swallows, the guffaws of ravens, the whistles of mockingbirds, and the lifting chirps of bobolinks, eastern meadowlarks, upland sandpipers, and all manners of other birds.

"Over increasingly large areas of the United States," wrote Rachel Carson in her seminal *Silent Spring* in 1962, "spring now comes unheralded by the return of the birds, and the early mornings are strangely silent where once they were filled with the beauty of bird song. This sudden silencing of the song of birds, this obliteration of the color and beauty and interest they lend to our world have come about swiftly, insidiously, and unnoticed."

The annual American report *State of the Birds* finds that "More than 97 percent of the native grasslands of the U.S. [including prairie and pasturelands] have been lost, mostly because of conversion to agriculture. As a result, grassland bird populations have declined from historic levels far more than any other group of birds."

Factory farms endanger not just farm animal welfare and human health but also the survival of wildlife.

CHAPTER FIFTEEN

AMERICAN MOUNTAIN MEADOW

THE BATTLE BETWEEN PASTORAL AND INDUSTRIAL

It is evident even from San Francisco's airport that Californians are environment- and health-conscious. Signs at security-check counters advise passengers to retain empty water bottles for reuse, and signs in restrooms encourage passengers to remember that paper towels come from trees. Yoga pants and mats are everywhere in San Francisco. (Even unathletic I, inspired by the fitness culture, went to long morning yoga classes that left me overbent.) But though the people of California are more interested than most Americans in ethical and healthful eating, California is, ironically, home to some of the largest factory farms in the country.

California dairies make New England dairies look like kittens compared to lions. In New England, a dairy with more than a thousand milking cows is considered large. In California (and also in New Mexico, Arizona, and Nevada), the *average* dairy has more than a thousand milking cows. At close to two million dairy cows, California's state herd is larger than Canada's, and is close to a quarter the total American herd of nine million dairy cows.

Signs at California dairies spell out warnings in both English and Spanish: DO NOT TRESPASS. PREMISES MONITORED BY VIDEO CAMERAS. ¡NO ENTRE! AREA DE BIOSEGURIDAD.

I went to the headquarters of Foster Farms Dairy, a company local to California. Its headquarters was an immense, maze-like complex of expansive office buildings and towering steel vats of milk. I parked my rented car at the back, close to the hundreds of milk tankers, and I spoke to a tanker driver with a heavy moustache. To my delight, he gave me a slip of paper with dairy-farm addresses.

Foster Farms has five dairies, each with anywhere from 700 to 1,600 milking cows. I drove to the largest one, located in Stanislaus County. There's one dairy cow for every three people in Stanislaus County, but an outsider would not sense this, for they are nowhere to be seen. There's also one Mexican person for every three residents in the county, and this is obvious immediately. I drove through neighborhoods of rundown Mexican restaurants and food stalls with names like Julia's Tacquería and Tacos Vallarta, feeling like I was back in Mexico.

I had no more than parked at the Foster Farms dairy when a tanned, squat man in blue coveralls appeared from behind me and started interrogating me.

"What do you want here?" he snapped.

"I would like a tour," I said.

"Why?"

"For my studies."

"What studies?"

"Farm studies."

"Do you want a tour in English or Spanish?"

"English."

"Not Spanish? You look Mexican."

"I'm not Mexican."

"Who told you about this farm?"

"A tanker driver at the head office."

"*Who*? *Which* tanker driver? I need a name."

"I don't have a name. . . . He had a moustache."

Scowling, acting like he was a storekeeper and I a thief, he led me into an office, which was a converted kitchen. I sat on a cracked, wobbling orange chair clutching my purse. Standing over me, glaring down at me, he called headquarters and repeated my story to verify it. He found no verification, since I'd never entered headquarters, as I'd already told him. "You can't have a tour of the farm," he told me after hanging up. "We used to do tours but we don't anymore because of mad cow disease, foot and mouth disease, and all the bad publicity."

I'd driven over two hours on California's packed highways, and I felt sick and sweaty, as I often did after my stress-filled driving. I was in no rush to return to the road, so I settled into the cracked orange chair and spoke to this man, called Eric. Notwithstanding his initial gruff demeanor, he turned out to be warm and well-meaning, and we spoke for two hours.

Eric was Portuguese and had moved to the U.S. at the age of fourteen. His title was Main Foreman, and he'd attained the position largely by being fluent in both English and Spanish. Bilingual foremen are a deliberate hiring policy at Foster Farms dairies because their English-Spanish fluency enables them to communicate with, and serve as a liaison between, white headquarters executives and Mexican farm workers.

Eric showed me his greatest pride, the 2002 county herd book. In 2002, Eric's dairy farm had received an award in the highest category, that of 750-plus milking cows. Eric and eleven other workers, all of them Mexican, were featured in a photo. Eric and I flipped through more herd books, ending with the 2011 county herd book, the most recent one he had. Dairy had changed dramatically in California in just nine years. In 2002, 750-plus milking cows had been the highest category in Stanislaus County. By 2011, much higher categories had been added: 826-plus cows, 1,126-plus cows, and even 1,761-plus cows. California dairies grow immensely year after year.

I asked Eric again if I could walk through Foster Farms' feedlots. "I'm sorry," he said. "If it was up to me, I'd let you. But I've been ordered to not let anyone in. If you want, you can call the big boss at the head office tomorrow morning and ask him. His name is Ronald."

Despite his warring coldness over the phone, big boss Ronald, to my great surprise, agreed to let me return for a tour of the Foster Farms dairy. He had one condition: he would be my guide.

Ronald considered strangers dangerous, and he thought himself best equipped to control and divert such dangers. As he strode toward me from his pickup truck, his greeting was: "This is going to sound crazy, but you're going to have to lock your purse in your car."

Nearing fifty, Ronald stood at medium height, had a lean frame, and wore black sunglasses and a white T-shirt tucked into blue jeans. He was the most intimidating and impersonal man I'd ever met—as indisposed to smiling as a robot, priding himself on a life lived without laughter. He was nevertheless a skilled salesman, a talented orator, speaking loudly and firmly in the manner of a preacher trying to convert the masses. His aim was to convert me not to any conventional god, but to his god: agribusiness.

"My dad started working for Foster Farms two months after I was born," he said. "I've grown up at Foster Farms. . . . I've been seeing an increasing split in the agriculture industry over time. On the one side is the organic-type farm—let's call it grassroots."

The term "grassroots" is right, I realized, both literally and metaphorically. Alternative farms tend to be more likely to rely on grass than corn, and they tend to be more independent, struck up in a "grassroots" sort of way, the non-corporates.

"I don't have a problem with grassroots farms, but when they say they're *better* than traditional farms—farms like ours—*that's* when I have a problem. . . . The media misrepresents things, like pink slime; the job they did with that—twisting the whole issue, getting into a crazy frenzy—it makes me angry. . . . The government has too many regulations. . . . The debate on cow hormones like BST is misunderstood. We should be able to give cows hormones to produce more milk, but the dairy industry is playing into the public's and the media's hands by being 'hormone-free,' by not using BST."

Recombinant bovine somatotropin, or BST, is an artificial, genetically engineered cow hormone that is injected into cows to amplify milk yield. It is banned in Canada, the European Union, Australia, New Zealand, and Japan because it increases the likelihood of cow lameness and mastitis, and it may also harm human health. BST is, however, legal in all fifty American states. The U.S. Food and Drug Administration approved BST in 1994 based largely on a study that tested its effects on thirty rats for ninety days. The study was done by Monsanto, which was at the time the producer of BST. Today, more than half of large American dairies use BST; one in six American dairy cows are injected with it regularly. The only way for an American consumer to be certain of not drinking BST is to buy organic milk, in which its use is forbidden.

"I'm cynical about BST," a pasture-based New England farmer had told me. "I look at it as another expensive input that a multinational drug company has come up with. BST puts more stress on the cows, and they don't last as long. It's not in farmers' best interests to use BST, but they're sold on the idea that they need to use it to compete. We don't need more milk; we already have an oversupply of milk. By using BST, dairy farms are producing more milk, but milk that people don't want to drink."

Ronald was content to converse in the parking lot. I asked him if we could instead walk through Foster Farms' feedlots. "No," he answered. Eventually tiring of the parking lot, though, he agreed, "Fine, I'll give you a quick tour and show you our cows. The first thing you should know about them is that they're pampered. We give them everything they need for their comfort."

The 1,600 milking cows were listless, sluggish giants lying in stalls. The total herd at the dairy, including young cows and calves, numbered upwards of 3,000.

Cow stalls were bedded with excrement, as I'd seen also in Vermont, but the excrement in California was very wet. Misters in the ceiling were sprinkling water upon the stalls, which meant that the cows lay in thick layers of old, dripping feces, soaking in the stink and bacteria.

The feedlots of Foster Farms were like the feedlots of Vermont, except that they were even longer and filthier. And yet the Foster Farms' website portrayed painted scenes of rolling green pastures sprinkled with cows. "For four generations," said the website, "the Foster Family has considered it a privilege to care for its dairy herds. Our nurtured cows are fed natural and healthy diets, provided quality time to graze upon open pasture. . . ."

There was no pasture; it was a lie.

There is only one respect in which California cows are better off than their counterparts in most other states: they are permitted to keep their tails. The state of California banned the practice of dairy cow tail-docking in 2009. The ban served as an impetus for the states of Ohio, Rhode Island, and New Jersey to also ban tail-docking. As such, the California ban shows that even local and state laws must be fought for tooth and nail, since they can have repercussions throughout the nation, and perhaps even the world. Throughout my agricultural trek in New England, dairy producers had spoken to me about the California ban, and some, including Nelson, had said they'd stopped their practice of tail-docking because of it.

Ronald and I strode past rows of hundreds of white hutches holding lonely, sullen calves. "We're giving the calves a good start in life," Ronald claimed.

Two Mexican workers, turning the corner, stopped in their tracks to see Ronald. "We treat our Hispanic workers very well," Ronald told me. He waved at them; they turned away and made off. "We have great conditions for them. At other dairy farms, they work fourteen to sixteen hours a day. Here, they work nine hours a day plus their lunch breaks."

In American agriculture, the generic term "Hispanic" is preferred to the more specific "Mexican." Noses, including Ronald's, wrinkled whenever I said "Mexican." A friend later told me: "Hispanic is a term we like because it makes it easier for us to lump them together like they're all the same."

Dairy workers are generally paid the minimum wage or an amount close to it. If they're illegal in the United States, and perhaps as many as half of them are, farms can get away with paying them even less than the minimum wage. Foster Farms managers sit down every week to compare the performance of workers on each farm, and also between farms. This environment of tense "internal competition" is fostered deliberately so that workers can be replaced as soon as their performance slips a notch. Workers are more replaceable than dairy cows. At Foster Farms, I could understand why a Mexican worker like Oscar, whom I'd met at Dubois in Vermont, was eager to return to Mexico.

Foster Farms, like other dairies I'd visited, was thinking of switching to a robotic operation. "Our second biggest cost, after feed, is labor," Ronald said. "We wouldn't need Hispanic milkers if we had robots."

Robotic dairies, which I saw firsthand at a farm show, are a new trend in the industry, forming a source of impassioned rumors and excitement both. The term "robot" derives from the fact that at such operations cows are milked not by machines managed by humans but by machines managed by themselves: robots. Robotic dairies have been taking off like rockets in Canada but have been catching on in the United States more slowly because of the bargain-basement price of Mexican labor.

The trouble with robotic dairies, to my mind, is not the bizarre technology itself, but the fact that such operations are devoid of both pastures and people. Dairy cows cannot step outdoors, and humans have no reason to step indoors. The closed circular system largely runs on itself.

Ronald and I stopped at a dark, stinking brown pond. It was a manure lagoon, thirty feet deep, with a crust on top embedded with bits of newspapers and plastic trash. California's dairy cows produce more waste than its human inhabitants. Across the United States, farm animals produce three times as much waste as American people.

In addition to its dairy feedlots, Foster Farms has a chicken-and-turkey arm as well, which is the largest chicken supplier on the west coast and among the largest in the country. Foster Farms' dairy and poultry brands are, however, kept entirely separate, with different names, packaging, and websites.

"No one told me this, but it's my opinion," Ronald said, "that the reason the Foster family decided to have two brands is that if anything happens to one brand, the other brand will still be safe. They don't want to put too many eggs in the same basket, so to speak. They have the Crystal brand for milk and the Foster brand for poultry."

In 2005, Foster Farms was awarded California's highest environmental honor, the Governor's Environmental and Economic Leadership Award. It was presented to the company by Governor Arnold Schwarzenegger for "exceptional leadership and notable contributions in conserving California's resources."

Looking at the languishing manure lagoon and the abundance of feces everywhere, I could not tell what resources Foster Farms was conserving. Also, Foster Farms was given the award despite the fact that, only seven years prior, in 1998, the company pled guilty to the discharge into a national wildlife refuge of 11 million gallons of storm water polluted with chicken manure. This environmental catastrophe set it in violation of the Endangered Species Act, for the manure killed endangered aquatic wildlife. Foster Farms paid half a million dollars in criminal fines.

Upon leaving Foster Farms dairy, I visited a Foster Farms chicken "ranch," even though Ronald advised me not to. The property consisted of a faded blue-gray house and eight long ramshackle sheds surrounded by a wire fence. Each shed probably held 20,000 chickens, but none of them could be seen. A Mexican man emerged from the house and stomped to my car. He yelled at me to leave, as his wife and three children peeked from behind curtainless windows. (A friend later offered me a reason for the virulent reaction: "They were nervous probably because they're illegal and they're afraid you're going to report them.")

As with dairy, the website of Foster Farms poultry is deceptive. In a video about animal welfare—a video in which there are no Mexicans to be seen—a Manager of Live Production states "A happy animal's a quality animal, and a quality animal is where we want to be."

Foster Farms urges consumers to join it in "supporting the humane treatment of all animals" by submitting a "Humane Pledge" online. Consumers are asked to agree with the following statements: "I believe that all animals—those in our homes, those in service to our country, and those raised on our farms and ranches—are entitled to humane treatment. I will live better by ensuring that animals in my care are treated with dignity and respect. I will shop smart by choosing poultry, meat, dairy and eggs that are humane-certified. I will choose humane and support producers committed to animal welfare."

But Foster Farms is not one iota different from any other factory farm. It may even be worse than most, seeing that its chicken products sickened more than 600 consumers with *Salmonella* in 2013 and 2014. In wake of the disease outbreak, the USDA tested Foster Farms chicken to find that one fifth of chicken parts were infected with *Salmonella*. The USDA threatened to close three of Foster Farms' facilities and forced a recall of its products.

Ronald was right—it *was* strategic of Foster Farms to have two business brands. Even as the reputation of the poultry brand became temporarily tarnished, the dairy brand remained safe.

Foster Farms obtained humane certification from the American Humane Association. Getting the "American Humane Certified" label in 2013 set Foster Farms in the company of several factory-farm corporations where undercover investigations have shown severe cruelty: turkey producer Butterball, American egg producers Cal-Maine Foods and Rose Acre Farms, and Canadian egg producer Burnbrae Farms. The "American Humane Certified" label unfortunately cannot be trusted.

My favorite part of my American expedition was an evening at a Vermont farm called North Hollow. My visit there was like an enjoyable movie rather than the usual macabre show.

I was greeted by four dogs and a young man. His name was Brad, he was twenty-one years old, and he wore a stained gray T-shirt and old khakis. He was skinny and sun-tanned, with an elfish nose and sandy hair. Of all the men I met in American agriculture, he was the youngest, smartest, most interesting, and most talkative—a breath of fresh air.

"I'm from North Carolina," he told me. "My dad's a cardiothoracic surgeon and my mom's a nurse. I go to Middlebury College and I'm interning at North Hollow Farm for the summer because I think we're going to get screwed pretty soon if we don't figure out our food system. I've been reading and learning a lot about food issues. My conclusion from everything is: everything you eat, you better watch out."

I asked Brad for a tour of North Hollow Farm. He asked me rapid-fire questions about myself, then pronounced, uncertainly, "Okay . . . let's start with the goats."

We got into his beat-up pickup truck and arrived at a rickety wood barn. Standing since the late 1800s, it was like a gaunt old man: bent and lopsided, but somehow still surviving, smiling. It was an orderly sort of mess, filled with hay and straw, barrels and shovels, and ladders to upstairs lofts. The gate outside the barn's yard was wide open—perhaps pushed open by the wind—and goats had streamed out onto the road. They were running about and eating the neighbors' bushes and grasses, bleating chaotically.

Brad, who'd never seen the goats escape before, jumped out of the truck and said to me: "Can you help me herd them? You're a godsend! You're a godsend!"

I agreed happily, thrilled to not be treated like a nuisance. Agile and athletic, Brad chased the goats and brought them one or two at a time to the fence gate, maneuvering them by their collars. Standing at the gate, I opened it for them to enter, but no sooner would they enter than one or two goats would push past from behind me and bound out. Brad would then chase them and bring them back in as well, before proceeding with the other goats.

Eventually, to my disappointment—for I was enjoying myself thoroughly—all the goats were returned to the yard of the barn and reunited with the menagerie of other animals who lived with them. These tenants were a flock of sheep, hens, ducks, miniature horses, and alpacas—most of them overseen by cats and kittens, who viewed them as sofa beds.

North Hollow Farm also had a dozen sows who lived in the downstairs portion of the barn. They had names like Pam and Betty, and they lived in spacious pens filled with hay and straw. Their piglets nuzzled one another and played together with high-pitched squeals. They enthusiastically chewed on my shoes, which I'd noticed that all piglets and pigs seemed to like to do.

As we left the barn, Brad turned to face me so suddenly that I bumped into him. "This farm may look bad but it's the cream of the crop," he said defensively. "When I came here, I thought things were screwed up. They were giving shots to the goats, and all the animals weren't outdoors. But now I see that North Hollow Farm is among the best there is. Most other farms are completely fucked up."

"I *know* that," I said. The strength of my response surprised both him and myself.

My mental journey had been similar to Brad's. I'd judged the first farm I'd visited, the Miller organic dairy farm, severely, but

American dairies really put things into perspective. The Miller farm in Canada had had sixty-five milking cows who were tied to stalls for two of three days of the year, but who grazed outdoors on the remaining one of three days. American feedlots, in contrast, have hundreds or thousands of cows who sleep in manure and who don't step out on grass for even one day in their lives. This is not to say that the Miller farm needed no improvement—it needed significant improvement—but it's to acknowledge that in comparison to the majority of its peers, it was better. Because I hadn't seen its peers at the time, however, I hadn't known that.

Brad and I climbed again into his pickup truck and he drove me to the main business of North Hollow Farm: beef cattle. He stopped at a wood barn that read MOUNTAIN MEADOWS, and we jumped out and peered over the verdant valley. It was vibrant with color and birdsong at sunset. Cows grazed in the tall grasses, looking like dots against the mountains.

"We have cattle on pastures throughout the mountain," Brad said. "Our total herd is four hundred to five hundred cattle at any one time of the year. North Hollow is a grass-only farm. That means our cattle eat only grass. A lot of farms finish their cattle on corn, but we don't do that. Grass grows free and naturally; corn doesn't. Some of the land we use is free, too. People on the mountain give us a free lease on pasture because our cattle prevent it from becoming brush."

As Brad and I drove back to my car, he pointed out more cattle in the dancing grasses. I found myself envious of him, of his youthful confidence, his idealism, his *joie de vivre*. I'd been like him once, but my vigor had vanished as a leaf flies away in the wind. I was only five years older than Brad, but after my international farm investigations, I felt a generation older than him. I could be his mother's age. Or his grandmother's. A part of me seemed dead.

"I think we should work without machines," I heard Brad telling me as we stepped out of his truck. "My ideal kind of farm is one in which two draft horses make everything work. . . ."

His words brought me to life like a heart defibrillator. I'd subconsciously been churning through everything I'd seen and learned over the course of my journey, but I hadn't crystallized it until now. All of a sudden, I saw that I had the answers. I made to Brad the first speech I'd ever made, a soliloquy.

"Technology's not a bad thing," I said. "It's a good thing. For the most part, it makes life easier and better. Your dad, a surgeon, probably uses a lot of technology in his work to improve patient lives. Technology can be used to improve the lives of farm animals or to enslave them. It can create comfort or eliminate it. It can build on nature or remove from it. The trouble is not technology; it's the use of technology.

"When it comes to agriculture, I'm realizing now that the primary divider in the industry is farm structure. In other words, whether the farm is pastoral or industrial. Pastoral farms are those where animals are able to exercise their nature—where they're able to walk and roam, to breathe freely, to feel sunlight. Industrial farms, in contrast, have crowded, barren, artificially lit, confinement environments. In every way, pastoral farms are better than industrial farms.

"If the primary split in agriculture is around farm structure, the secondary split is around farm size. The majority of animal farms are either small or else quite large. Of the dozen dairies I visited in the U.S., each had less than 120 milking cows or else more than 480 cows. There was nothing in between. This difference is a magnitude of four times. It's like a huge chasm in the center of the industry.

"The reason for the gap is that mid-sized dairies grow to become large dairies, and small dairies sell off to large dairies. As such, the ranks of large dairies continue to grow every year, and the ranks of small and mid-sized dairies continue to slim. I saw an awful dairy farm in Vermont called Dubois, which has quadrupled in size from 300 milking cows to 1,250 in just one decade. The size split in agriculture is becoming more pronounced over time.

"Pastoral farms tend to be small, and industrial farms tend to be large, but this is not a definite rule. There are also industrial farms that are small, and there are pastoral farms that are large. We can think of farms as falling into one of four sections of a matrix. The sections of the matrix can be called: large-industrial, small-industrial, small-pastoral, and large-pastoral.

"Large-industrial means a large factory farm, like the dairy feedlots I've been seeing in the U.S. and like the chicken, egg, and pig farms I've seen around the world—in Canada, Malaysia, Mexico. Large-industrial operations tend to benefit from economies of scale but at the cost of animal suffering. Large-industrial easily forms the largest section of the farm matrix today.

"Small-industrial means a small factory farm. For instance, an egg-laying hen farm I saw in Belize had a thousand hens—a very low number compared to others—but it housed the hens in wire cages. Small-industrial farms are few because they don't make sense. They're too small to have any economies of scale, and their animals have miserable lives. They're a lose-lose situation for all, and that's why they form the smallest section of the farm matrix.

"Small-pastoral is the oldest sort of farm in the world. In the past, all farms were small-pastoral, raising a few cows and pigs, and a flock of crowing roosters and hens, on yards or small pastures attached to the house. In the present day, only a small minority of farms around the world remain small-pastoral, mostly in developing countries and traditional communities. I saw several in the villages of Indonesia. The best small-pastoral farm I saw was in Belize, a Mennonite women's farm. Small-pastoral farms tend to treat their animals well, but just like any other small business, they tend to have a tough time financially because they don't have economies of scale. That's why their ranks drop every year.

"The last and latest section of the matrix is large-pastoral. A large-pastoral farm is a place like North Hollow Farm. On about a thousand acres of mountain pastures, you have four hundred to five hundred beef cattle, a hundred pigs, eighty goats, and some sheep and chickens. Another large-pastoral farm I've visited is in Canada, owned by an English farmer called Roger Harley. His farm also measures about a thousand acres and has hundreds of sheep, pigs, and cows who live outdoors year-round. Your farm and Roger's farm are large and pastoral: they're large-pastoral.

"Large-industrial and large-pastoral farms have nothing in common except size, and even that, to a limited degree. Large-industrial operations keep the same kind of animal in high numbers—65,000 egg-laying hens under one roof, for instance, as I saw in Malaysia. Large-pastoral farms tend to have different kinds of animals because they depend on natural cycles for their sustainable continuation year over year. Even at their largest size, large-pastoral farms never become monstrous, because it would become more difficult to maintain ethical standards and ecological balances.

"The farm matrix is helpful because instead of viewing agriculture as an incomprehensible maze, we can start to think of it as a house with four doors. Every farm enters and exits through one of the four doors. The large-industrial door may be the most obvious, immense door, located at the front of the house, but it is not the only door. The most welcoming door is the newest one, the large-pastoral door. Farmers can, if they so choose, exit the large-industrial door and enter the large-pastoral one. If they do so, they will find a way of working that suits their conscience instead of scratching it out.

"Large-pastoral farms have economies of scale, and so they fulfill the preferences of the producer (low cost) and the consumer (low price). Importantly, they also have animal welfare and they're sustainable. Large-pastoral farms stand alone in the matrix as being a win-win scenario for all involved. We're never going to return to a time of small-pastoral farms, but we *can* create a time of large-pastoral farms."

As I said the next words, my heart seemed to almost stop, for I'd waited so long and worked so hard to be able to say them. I'd searched all the corners of the globe for solutions, and, here, atop a mountain in Vermont, looking out over lush valleys, I'd *found* a key solution. The moment was so perfect that it seemed too good to be true. "An important solution," I uttered, "is large-scale pastoral agriculture."

I peered nervously into Brad's face to see what he would say. He'd been listening with patient concentration throughout my sermon on the mount, his arms crossed over his chest.

"Send me your full report when you're done," he pronounced.

Others may have been disappointed, but to me, this was the ideal response. *Interest.*

CONCLUSION

THE FUTURE IS ALREADY HERE

A SINGLE GARMENT OF DESTINY

I climbed the agribusiness ladder in not one but eight countries of the world: the United States and Canada, Mexico and Belize, Indonesia and Malaysia, and Singapore and the United Arab Emirates.

Along the way, I met all kinds of people, ranging from contract growers to chief executive officers, farmers to slaughter workers, veterinarians to government officials. I investigated all kinds of farms: large-industrial, small-industrial, large-pastoral, and small-pastoral, per my new farm matrix. I saw all kinds of farm animals: dairy cows, beef cattle, veal calves, sheep, goats, pigs, egg-laying hens, chickens, and turkeys.

If I were to sum up everything into a single sentence, it would be a sentence by, appropriately, a science fiction writer. "The future is already here," said William Gibson, "it's just not very evenly distributed."

When I visited pastoral farms, I found that everything was obvious, since it is only a trinity of elements that are involved: sun, grass, animal. Industrial farms, in contrast, are confoundingly unintuitive, as one deprivation leads to another deprivation, and that deprivation leads to the next deprivation, and the tangled, intertwined chain of deprivations requires considerable explanation.

I found that the pattern of factory farming everywhere follows the trajectory of the United States. Chickens and egg-laying hens are confined and commoditized first, followed by pigs, then by cows. It's as if the industry starts with the animals it considers least sentient first, then works its way up the ladder.

The U.S. Environmental Protection Agency has invented a term for factory farms: concentrated animal feeding operation (CAFO)—because large numbers of animals are concentrated and fed at a single site. Originating in the United States, CAFOs are growing in both size and number around the world. The majority of the world's chickens, egg-laying hens, and pigs today live confined to industrial factories.

Most farm animals today live and die without ever feeling a ray of sunshine on their backs or a blade of grass under their feet. Every hour builds upon the next in a perpetual hell, and the misery continues onward without reprieve. Existing in conditions of disease and decay, animals die before their deaths.

Farms are the modern *1984*. Animals are kept in the dark, and people are also kept in the dark. Official party lines are disseminated to the public and readily accepted by it. At every meal, people swallow small and large doses of lies.

"Things like the continuance of British rule in India, the Russian purges and deportations, the dropping of the atom bombs on Japan, can indeed be defended," wrote George Orwell, "but only by arguments which are too brutal for most people to face."

Industrial agriculture can also be defended, but only by arguments that are too brutal for most people to face. When we argue that saving a few cents on an egg is more important to us than the welfare of the hen who laid the egg, what we are saying is that our minor, trivial interests deserve greater consideration than the major, welfare interests of a living being. We are stating that we are willing to condone anything—and everything—so long as our wallet remains stiffly uninvolved. Such thinking diminishes what makes us human.

In *Animal Machines*, a book from the 1960s about factory farming in the United Kingdom, Ruth Harrison wrote: "The arguments against factory farming are essentially based on humanitarianism and quality; the arguments for factory farming, such as they are, are economic arguments. We need not be disturbed about this," she continued optimistically. "Throughout our society there are clashes between economic and social considerations; this has been true from the first factory legislation onwards. Our legislation is full of examples of laws which for humanitarian reasons prevent people using the cheapest methods."

Industrial agriculture is like a highway with no street signs, no speed limits, street lights, lane markings, or laws. Cars and trucks barrel onward haphazardly, often colliding with one another, risking not just a multi-vehicle pile-up that would bring everything to a grinding halt, but also injury to the bystanders going about their lives along the sides of the road.

Agriculture is a "free for all" industry, which ironically makes it costly and dangerous for all. There is no one in charge of the expanding anatomy of animal agriculture. The bloated body, one finds upon examination, is devoid of a mind.

The agriculture industry claims to be strong and healthy, but upon inspection one finds that it is not a vigorous young man bursting with life, but a paranoid and senile old man who lives in terror that any encounter with outsiders will spell his death. He shuts down completely, closing all doors and windows, forbidding all contact with the external world, surrounding himself with barriers of "biosecurity." He fails to realize, however, that no sustainable path to the future can be paved with fear. The only roads paved with fear are those that are at their end.

"Injustice anywhere is a threat to justice everywhere," wrote Martin Luther King, Jr., half a century ago. "We are caught in an inescapable network of mutuality, tied in a single garment of destiny. Whatever affects one directly, affects all indirectly."

When animals suffer, people also suffer. Many farms I investigated were in the throes of disease. Farmers throughout the world expressed to me fears that an outbreak might suddenly sweep across their operations and consume their animals. My trek through Southeast Asia especially showed me how damaging a bird flu outbreak can be—to both people and animals—having repercussions even a decade later. Every year, newspapers around the world contain bold, panic-ridden headlines of new variants of diseases that have originated in the bowels of some chicken, pig, or cow factory, and that threaten crossing over to humans.

There are bird flus and swine flus—recently, H5N1, H7N9, and H1N1. There are variants of *Salmonella* and *E. coli*. There are other farm diseases, too—Newcastle disease, infectious bursal disease, ringworm, mastitis.

A new deadly disease has also emerged in 2013, called porcine epidemic diarrhea virus (PEDv). The virus, which spreads through manure, causes rapid diarrhea, vomiting, and dehydration in newborn piglets, and kills a startling 50 to 100 percent of all those affected under seven days of age. Thousands of pig operations across the United States and Canada have been infected with PEDv and many millions of piglets have perished, their palm-sized bodies oozing with suffering. At Charlie's pig factory farm in Canada, I'd been horrified to see pails of dead piglets lining the corridors; today, I cannot imagine how many pails there must be.

Despite the preponderance of factory farms, it is uplifting to note that there is no reason for the most brutal systems in our world today to exist. There is no decree anywhere that declares that the survival of the human race depends on the caging of egg-laying hens, the crating of sows, or the banishing of sunshine on farms. These methods are the outcome not of necessity, but a short-sighted mentality. They can be eliminated just as decisively as they have been erected.

The fact that there are solutions at all to animal agriculture is a cause for celebration, for there are many areas of life that are absent of clear answers. "Sometimes the questions are complicated and the answers are simple," wrote Dr. Seuss. Such is the story of animal agriculture.

Solutions can be divided into two categories: production and consumption, relating to producers and consumers.

When talking about producers, it is important to separate the people from the problem. My cultural immersion in the deepest, thickest veins of the industry globally showed me that people everywhere are, for the most part, good. Even in those operations that I could not condone, I came to admire and appreciate the people I met. In addition to educating me about their industry, they educated me about life.

Brick Roberts, for instance, taught me the importance of consuming wisely, of minimizing my environmental footprint. Jim, my veal farm driver, showed me that it is possible to be happy despite life's tragedies. Slaughter worker Nader, whom I met at Blackwater, showed me through his anguished mental state how averse the human heart is to violence—how the knife turns inevitably inward. Agung, my spiritual strawberry farmer friend from Indonesia, taught me to cultivate peace of mind.

It's not farmers and workers who are the problem in agriculture, but the broader system that permits them no room to pause and think. "So often the problem is in the system, not in the people," wrote Stephen Covey in *The Seven Habits of Highly Effective People*. "If you put good people in bad systems, you get bad results."

"The real ethical issue about factory farming's treatment of animals," agree Peter Singer and Jim Mason in *The Ethics of What We Eat*, "isn't whether the producers are good or bad guys, but that the system seems to recognize animal suffering only when it interferes with profitability."

I propose eight producer solutions to improve animal agriculture: large-scale pastoral farms, natural breeds, gender diversity, internal commitments, meaningful inspections, decisive lawmaking, accurate marketing, and organic strengthening.

"The question in agriculture today," Roger Harley in Canada told me, "isn't whether there should be scale of economy—there *should* be scale of economy—but whether that scale should be supplied by a natural outdoor farm or a factory farm."

I hope that the best examples of large-pastoral farms I have seen—Roger Harley's farm and North Hollow Farm of Vermont—can serve as role models in a chaotic, confused industry that is in dire need of them. Large-pastoral farms uniquely meet the needs of animal welfare, farmer livelihood, consumer prices, and environmental sustainability.

The world must not be perceived as an apple pie where, if we cut a larger sliver for animals, there remains a smaller slice for us. After all, it's not just animals who benefit from better living conditions at farms. It's also everyone else, including consumers, producers, and the planet itself. Consumers benefit from increased food safety, farmers and workers benefit from a better work environment, and the earth benefits from less degradation. All of these are goals worth fighting for.

A second area of concern is genetic selection. Today's farm animal breeds, especially those of chickens and turkeys, are extremely unnatural—their heart and legs unable to keep pace with the rapid expansion of their flesh. Narrow breeding goals also lead to aggression in animals. When egg-laying hens peck and kill one another, when pigs bite each other's tails until they bleed, we know that something has gone dreadfully wrong. The experiment has failed. A reversion to natural breeds would mean animals with healthier body structure and behavior.

Third, gender diversity is something that no one likes to talk about, but that everybody notices. The number of women in agriculture is like a bald spot on the head. A few sparse hairs grow there, but everybody knows that the head is bald. Data across industries suggests that gender diversity improves decision-making. My own experiences and analysis show that the hiring of women would be beneficial for animal agriculture, ushering in greater concern for animal well-being and food safety.

A fourth recommendation is the industry's committing to progress of its own volition. For instance, pig producers in Europe have created a *European Declaration on Alternatives to Surgical Castration of Pigs*, and the American Veal Association has called for U.S. veal farms to transition from individual calf crates to group pens. Occasionally, agribusiness commitments are political ploys—mentioning either no dates or else far-off dates that get pushed even further as they draw closer—but at other times, they are sincere and effective. They do not have the regulatory power of laws, but they set internal aims and directions.

A fifth solution is inspections. Farms are not inspected by the government at all—and should be. As for slaughter, inspectors in the United States and Canada are friendly and chummy with industry, a relationship that defies the very purpose of inspections in the first place. Slaughter-inspector incentives are in dire need of a makeover. Today, a makeover is occurring, but in the wrong direction. Both the U.S. and Canada are working on privatizing slaughter inspections, such that killing would be inspected not by government employees but by company employees. The ostensible purpose is to increase slaughter line speeds, a move that would all but assure even more animal cruelty and less worker and food safety. Self-policing at slaughter is a case of the fox guarding the henhouse.

A sixth and related avenue for change is regulations. Farm animal welfare standards are necessary. At the least, animals should be recognized under the law as what they are—animals. European law recognizes animals as "sentient beings," and requires nations to, when formulating and implementing their policies, pay "full regard to the welfare requirements of animals." Animal welfare in Europe is legally on an equal footing with other democratic principles like gender equality and human health. In the United States and Canada, in contrast, there is no law that distinguishes a pig from a table. Both have four legs, and both constitute forms of property.

Battery hen cages, gestation sow crates, and veal calf crates are banned throughout the European Union. In contrast, the first two of these extreme forms of confinement form the *modus operandi* at American and Canadian farms. In the U.S., laws protecting farm animals are minimal, and, where they exist, are on a state level rather than federal. But while a minority of American states are making progress, others are degenerating by passing ag-gag laws that, by preventing documentation, prevent an impetus for the passage of welfare laws in the first place. Moreover, state laws known as "Common Farming Exceptions" allow industry—rather than lawmakers—to make any practice legal so long as it's common.

Seventh, in addition to laws that protect farm animals, what is needed is regulation to ensure accurate marketing. Labels on meat, egg, and milk products often mean little to nothing. This is a deliberate strategy on the part of agribusiness.

"It is almost universally felt that when we call a country democratic," wrote George Orwell, "we are praising it: consequently the defenders of every kind of regime claim that it is a democracy, and fear that they might have to stop using that word if it were tied down to any one meaning."

Labels like "fresh" and "natural" are not tied to any one meaning—hence their prevalence. The term "naturally raised" means too little in the U.S., and too much in Canada, where its use is hardly permitted. The description "raised without hormones" is often used for chickens, turkeys, and pigs, but it is deceptive because hormones are not generally used in these animals—they are used in dairy cows and beef cattle. The description "raised without antibiotics," in contrast, is useful, since antibiotics are widely used in farm animals, causing antibiotic resistance.

Some labels relate to farm animal diets. The term "grain-fed" is getting increasing currency these days as a promotional tool, but it is usually equivalent to "corn-fed," and thus has no value as a differentiator since corn is the standard diet at animal farms. The term "vegetarian-fed" means that farm animals have not been fed slaughter by-products, and the term "omega-3" means that they have been fed foods rich in omega-3 (hopefully flax seeds instead of fish). These two labels have their uses, but they say nothing about living conditions.

With regard to living conditions, the term "free-run" means that animals have been permitted to roam indoors. Free-run eggs are always better than eggs from cages, but the label means little in other areas where it is increasingly being applied, such as broiler chickens and turkeys, who are not kept in cages to begin with in the United States and Canada. The term "free-range" is superior to "free-run" as it means that animals have outdoor access, although the level of outdoor access can vary considerably. Both "free-run" and "free-range" are in need of rigorous defining and tightening.

Finally, the granddaddy of labels, the most serious and weighty of them, the one with the highest potential, is organic. "Organic" consists of a range of stipulations that are stringent in areas like pesticides and drugs, but are insufficient in other areas. Concrete lots instead of pasture, small doors leading to the outside, castration, artificial insemination, and crowding are today the norm at many organic farms. The minimum outdoor requirement at organic operations in the United States and Canada is 120 days a year—one of every three days—but it should be increased to at least 180 days, and preferably 240 days.

The best farms I have visited around the world have not been organic. Some of the organic farms I have visited with lofty expectation have been letdowns. This is because organic methods are often detached from the intentioned spirit of the label. The *spirit* of organic is to be natural, ethical, and biological—the French translation of the word is *biologique*. The reality of the label is often procedural, profiteering, and bureaucratic—checking off long forms, filing stacks of papers, and counting wads of cash.

Organic has important potential as a differentiator between farms, but it has yet to reach that potential. It is like a house in the process of construction. The foundations are sturdy but the fixtures are shoddy and risk damaging all the carefully laid groundwork. Organic must be strengthened in order to be meaningful.

In my experience, those farms that start out as organic—committed to the philosophy from the very beginning—are usually more ethical and ecological than those that, upon sniffing the air of business opportunity, decide to convert to organic. (Or, in the case of corporations, decide to offer lines of organic products.) The reason is that when farms convert, they have a powerful cost-incentive to keep everything the same. As such, the conversion often occurs more in label than fact. The product gets re-priced, the packaging gets revolutionized, but animal lives remain mostly the same.

My eight recommendations—relating to pastoralism, breeds, gender, commitments, inspections, regulations, marketing, and organic—offer steps in the right direction. The cover of this book represents the ideal of my vision.

"How can you talk about only ignorance?" wrote Anna Sewell in *Black Beauty*. "Don't you know that it is the worst thing in the world, next to wickedness?"

My investigations helped me jettison my own ignorance. My writing about my investigations has been an attempt to jettison the ignorance of others. Some people assume that animal factories do not exist because they have never been to them. This is like assuming that prisons do not exist because one has never been to them.

My project has focused less on a renunciation of meat and more on a revaluation of the excruciating methods that are used to produce it. That said, a note must also be made about the consumption of meat. The number of animals we eat around the world defies all reason, and yet it is increasing. We as consumers often know what we want, but we don't know the *consequences* of what we want. Low prices of meat, milk, and eggs are achieved only at steep costs—costs that require us to rethink our decisions.

"Eating industrial meat takes an almost heroic act of not knowing or, now, forgetting," states Michael Pollan.

"Maybe, in the grand scheme of things, the life of a pig or cow or fowl of the air isn't worth much," writes Matthew Scully in *Dominion*. "But if it's the Grand Scheme we are going by, just what is a plate of bacon or veal worth? . . . If animals are just commodities, then we are just consumers, with no greater good than material pleasure and no higher law than appetite."

Most of us today eat animals because we want to, not because we have to. But there are too many of us in the world to be eating as many animals as we do, and there is too much animal suffering for us to be eating them as guilelessly as we do. Large-pastoral agriculture is a possibility for the planet if meat consumption decreases simultaneously. Otherwise, the numbers of animals themselves hint at a factory system.

There are several ways to accomplish a reduction in meat consumption. One method is to eliminate meat from the diet, becoming vegetarian or vegan. Another popular, easy global initiative is Meatless Mondays, in which people pledge to eat plant-based diets on Mondays. A third option is to reduce meat consumption to one meal of the day. (Mark Bittman explains such a concept in his book *VB6: Eat Vegan Before 6:00 to Lose Weight and Restore Your Health . . . for Good.*) Whether removing animals from the plate entirely, or removing them on certain days of the week or for certain meals of the day—the goal of eating fewer animals can be achieved in multiple, manageable ways.

It's erroneous to think that it's only animal welfare that's at stake every time we sit down at the dining table. It's also human health and habitat. Countless studies today demonstrate that reducing animals in our diet, and replacing them with fruits and vegetables, is good for our health.

Also, factory farms harm the environment. Forests are flattened to grow corn to feed to farm animals. Globally, tens of thousands of acres of forests are lost every day—an area twice the size of Paris—summing up to millions of hectares a year. Moreover, farm animals gain in weight only a fraction of the weight of their food and excrete the rest. The copious quantities of manure pollute land and water.

"When we try to pick out anything by itself," wrote John Muir, "we find it hitched to everything else in the universe." Such is the story of every cow, pig, and chicken on the planet.

As of the date of publication of this book, I have not eaten animals for ten years, and lived to tell about it. Hindus in India have had a vegetarian heritage for thousands of years, and there are hundreds of millions of them to tell about it.

At the very least, I recommend shopping carefully. Here are some tips and guidelines.

Throughout my expedition, I was shocked to visit awful farms and then to visit their websites—and find that the two have nothing in common.

Some websites look like vacation destination pages. They display arrays of stunning pictures and colorful cartoon images. Their language is lyrical and sometimes strangely sentimental, the words seemingly composed by a poet. They profess great animal care and consideration, and even go so far as to talk about "humane pledges" or to quote Gandhi. They assert that not only are they willing shepherds of animals, but also stewards of trees. At times, they sound more impassioned than PETA or Greenpeace.

But none of this is new. "Assurances that all is well in the world of farm animals have not been lacking in the immediate past and will certainly not be lacking in the future," wrote Ruth Harrison in *Animal Machines* long ago. "We shall be assured that no cruelty is involved in intensive rearing. . . . We shall be assured that the products of the industry are better and more nutritious than they have ever been and we shall be told that we are the best fed people on earth and that we are becoming better fed every day. It will be imputed that those with a sneaking suspicion that all is not well are a decided minority."

Now for the recommendations: start shopping at farmers' markets and specialty stores. Farmers' markets and specialty stores pre-select farms for customers, using measures like animal welfare and environmental sustainability. At a farmers' market, you will have your choice of small-pastoral products (and perhaps also some large-pastoral ones). At an established specialty store like Whole Foods, you will have your choice of large-pastoral products (and perhaps also some small-pastoral ones).

All products at a farmers' market or specialty store will not necessarily be humane, but you can expect that a much greater proportion will be than at a conventional retailer. At any rate, you will be able to ask questions and find out.

If you do not always have the time or the money to seek out an alternative retail chain (I often don't), you can still do well. When you go grocery shopping, ignore all the grassy pictures on the product packaging. Pay attention only to labels. Many labels are a quagmire of issues, as discussed, wanting for more regulation, but others offer a useful form of differentiation.

Here's a primer of labels: 1. Most meaningful labels: organic, free-run, and free-range. 2. Somewhat meaningful labels: raised without antibiotics, raised without hormones (when on milk and beef), and vegetarian-fed. 3. Meaningless labels: fresh, natural, naturally raised, and grain-fed.

When you return home, visit the websites of the companies you saw in the grocery store. You will often find that the larger, more common brands, which you thought were unconnected, belong to the same company. Ignore all cartoon images online. Ignore all photos that look like postcards. Ignore all photos in which no animals are present—a red brick house, a sprawling tree, a shimmering lake, a green mountain. Ignore all photos of a man in overalls petting a calf, of a giggling girl collecting an egg, of a grinning boy holding a piglet. Ignore promotional claims like "family farmer" or "third generation."

On the website, go to the "Contact" page and call or e-mail the company with questions. Ask them whether they permit public visits (this is a good test, whether or not you actually intend to visit). Then ask specific questions about how animals are treated. Some good questions to choose from are: How much space are animals allotted? How much time (if any) do they spend outdoors? Are any sorts of mutilations done (castration, tail-docking, de-beaking, declawing)? Are animals regularly given doses of antibiotics or other drugs? With regard to specific animals: Are egg-laying hens confined to cages? Are sows restricted to crates? Are dairy cows chained to stalls? Do chickens and turkeys belong to industrial, unnaturally fast-growing breeds?

If you do not receive a reply, you will have your reply.

This legwork is easier than it looks, because once you find a company you're comfortable with, you can stay with them.

With every visit to a cash counter, we, consumers, cast votes for certain values over others. In democracies, politicians are beholden to their voters; in free-market economies, companies are beholden to their customers. It's we who have immeasurable power, if only we choose to wield it. We must have opinions and we must express them loud and clear.

EPILOGUE

THE END OF EXPLORING

THE APPLE OF TRUTH

"Darkness has no existence of its own," I once read. "It's only a lack of light."

My hope is for this project to serve as a lantern in a night sky. It has been mine, but it has also belonged to all the men and women whom I met along my long, winding way. As we arrive together at the end, I would like to offer an update on everyone.

Organic dairy farmers Michael and Irene Miller continue to wage war with each other but remain together. Irene's dream of leaving the dairy farm never came true. Instead, she has, sadly, developed early-onset Alzheimer's, and suffers memory loss, recognizing close family but not friends or acquaintances. Michael and Irene's daughter Annie continue to work at the farm. The Millers have installed a robotic milking system. (Their farm is now organic and robotic—two words that should form an oxymoron.)

When I'd stayed with Brick Roberts, I'd hesitated to call him Uncle Brick, even though I knew he preferred it. Now, at the end of my journey, I find that I can call him Uncle Brick with ease. Even if Uncle Brick and I were not to see each other for twenty years, we would recognize each other's voices upon a word, and would enjoy the same level of familiarity we did during our first conversation. Talking to gruff, cheerful Uncle Brick is like warming my hands at a fireplace—it is both electrifying and calming.

Uncle Brick and Aunt Jane are doing well, as are their children. Nick, their elder son, married a nurse and has a little daughter who brings her parents, grandparents, and Uncle Paul great joy. Uncle Paul, who remains Uncle Brick's best friend, has not been faring well. He's suffered a stroke and is in recovery. Nick's younger brother Will continues to be "a ladies' man," Uncle Brick says. "There's always several young ladies showing up here—sometimes at the same time," says Uncle Brick, laughing. "Will is tasting the dragon before he bites."

The Roberts family's beloved dog and cat have died of old age, and the dog has been replaced by another, who "loves kids and generally makes a great big nuisance of herself." Uncle Brick is still "producing cheap food," as he himself describes it. His egg-laying hens continue to live in dreary cages, but the broken fence outside the turkey shed has been fixed, and the free-range turkeys "like running around outside, doing what they do."

Uncle Brick meanwhile travels in North America and Europe to lecture about "sustainable farming, good husbandry for animals, and solar energy recovery. We've won some very prestigious awards, ya know." (No comment.)

Nick and Will's friend Charlie continues to grow pigs on contract. He and his wife now have two children, who often play in sandboxes with Nick's daughter. Charlie's pig factory is not doing well, however. "They're having a very tough time of it," Uncle Brick reports. "There's a pig disease around, a virus. When your pigs get that, your mortality gets really high. Death losses are ninety percent. There's no resistance to it. It's kind of a superbug. Piglets scour, and then they dehydrate, and then they die on ya. There's about four big diseases in the pig industry today. It's terrible."

Nick's friend Terry, who is a chicken grower, has also gotten married. He suffered a car accident, however, when his four-wheeler hit a deer. "Once you get hit with that magnitude," Uncle Brick says, "you're never the same. But Terry's getting better. Some broken bones, but nothing too permanent, I think."

I paid a visit to Roger Harley and his family. It was Roger who'd first planted the idea of "large-pastoral" in my mind, through the example of his own farm, and it was Roger who'd shown me that things can be different. Previously, I'd visited the Harleys in winter and spring; now I went in late summer, when the feeling at the farm was quiet, hot, and energetic.

When I'd first met them, the Harleys had had 300 farm animals. Today, they raise 1,500 pigs, sheep, and cows, and their herds continue to grow every year. Their lands are vast enough that even these animals seem to disappear among all the grasses, slopes, and trees.

The Harleys' five herd- and guard-dogs followed the five of us—Roger, Julie, Emily, me, and my husband—as we trekked about. Every auburn sow, along with her piglets, was allotted an immense pen furnished with a roomy wood house and a muddy wallow for herself and her piglets. Black-and-white cows and curly-haired sheep and lambs grazed on hills. Roger and Julie Harley's children, James and Emily, are now young adults, and express enthusiasm about continuing their parents' pastoral ideals into the future.

Roger and Julie Harley attended my wedding. On this note, I finally stopped being single. Soon after my travels ended, I met someone. After months of dating, we decided to get engaged. After months of engagement, we were married. The Harley farm was the first that my husband ever visited; he liked it.

Jim, the elderly man who'd driven me to the veal farm, is in poor health, but his personal life has come full circle to a happy ending. "I try to stay out as much as I can. I take a drive to relieve my aches and pains. Today, I'm goin' to go see a veteran who's ninety-two years old. Ya know I like supportin' the troops. I've let my house go to hell after losin' my sister and niece. Goin' to that farm with ya was a sort of therapy. . . . I'm gettin' real old. I'm gettin' X-rayed from top to bottom these days. My diabetes is gettin' worse, too. I'm takin' a bunch of pills. . . . I have a younger brother. I never mentioned him to ya because he and I were estranged from one another. We had a fight long ago. We finally got together after twenty or twenty-five years. I'm very grateful he found me. I'm a stubborn bugger when it comes to such things, sweetheart."

The horrific slaughter plant Blackwater, I am relieved to note, closed its doors. It was not shut down by the government; rather, the owner, Abdul, decided to return to his home country, Pakistan, for personal reasons. I was unfortunately unable to get in touch with slaughter worker Nader, whom I'd come to know that day. I imagine he continues to toil on some kill floor somewhere, changing his residence, workplace, and cell phone number as he evades the authorities, so no one learns that he has a measly income outside of his disability insurance.

I miss Agung, my spiritual Indonesian friend. The Shiva statue he gifted me from his shrine sits on my bookshelf, reminding me of his peace and wisdom. It will remain on my bookshelf always. Agung's strawberry farm, his pride and joy, is managed today by his younger brother Ketut. Were Agung alive today, my husband and I would have honeymooned in Bali, as I'd promised Agung.

The Mennonite women of Belize are thriving—mother Geraldine Lemon, daughters Abbey and Nancy, and granddaughter Katie. They are pleased to hear that I am now married, as that had been a special source of concern for Geraldine, who'd viewed me as a wayward hippie. Abbey is just as charming and cheerful as ever.

"I wish for you a wonderful life with your new husband as well as a dozen children to keep you young," she told me. "We're all doing great. Everything here's the same as before—we have our small herd of twenty cows and our little flock of hens. The cows have grown bigger since you saw them. I manage all the farming. It's just us girls here still! We're still selling weekly at the farmers' market. In Belize, I'm glad there are no franchises like McDonald's and Burger King!"

In real life, there is rarely a happy ending. In my case, I am pleased to note that there is. Everything led *somewhere*. I succeeded in entering an extremely secretive industry, and in documenting what I saw, smelled, and heard while I was inside.

My story began as a volunteer vacation at an organic farm, and became, without my knowing it, a global expedition into the deepest, darkest recesses of the international animal agriculture industry. Nothing about my story was easy.

At times, I felt as though I'd leaned forward to bite an apple but had bitten the entire tree. At every turn, I was challenged emotionally and ethically. I was constantly called upon to plunge out of my comfort zone. I had to put my life on hold, both personally and professionally. I was like Alice in Wonderland who plunged down the rabbit hole, floating from farm to farm around the globe like a loose and wild cannonball. I was more tested than I'd ever been, to the extent that I began to remember my midnights on Wall Street as the easy, idle days.

My descent from the midtown Manhattan high-rise cubicle that I'd occupied showed me that the world is more complex and confounding than I'd ever imagined it to be. I no longer entered brightly lit offices through smooth, sliding front doors; I entered dark factories through DO NOT ENTER WITHOUT AUTHORIZATION doors. I no longer wore high heels and dry-cleaned suits, but stained pants and cheap boots. I no longer strode on marble, but on manure. I was no longer a swallow—sweet, singing, soaring—but a rat—dirty, digging, detecting. From my deflated, depreciated position on the ground, everything looked different than it had from high up in the sky.

At times during my expedition, I felt like I was fighting through a fog, and the more I thrashed, the more blurred became my vision. The project that became this book took, from conception to publication, more than four years and thousands of hours of diligent, incessant effort. Like a marathon, it was uplifting but exhausting, nourishing but depleting. For a time, I felt as though instead of trying to free animals from their prisons, I was succeeding only in imprisoning myself. I felt as though it was not I who had selected my subject but my subject that had selected me, and now, like a bruising rope around my ankle, was refusing to let me go.

I often came close to giving up, to packing up my bags and returning to a normal, happy life. But something would stay my hand, would force me to keep going until I'd found the answers to the problems I was seeing. Once my eyes were opened, I found that I could not close them again, despite my wanting to sleep it all off and wake up pretending it had been a nightmare.

The only reason I persevered to the end, I realize now, is the sheer, stupendous number of farm animal lives at stake. More than 70 billion animals are reared for food each year around the world. We live on a planet on which there are—at this very moment in time—three chickens for every human being. Surely, on account of their numbers alone, farm animals must be accorded some importance in our moral compass.

Experience is never complete. It is an immense sensibility, a kind of huge spider web, woven of silken threads, suspended in the chamber of consciousness and catching passing particles in its sticky tissue. I have attempted, in the research of this book, to be a collection of senses. I have attempted, in the writing of this book, to be a windowpane—a sheet of glass for people to peer through. The word "essay" originates from the French verb *essayer*—to try. This project has tried to understand the roaring beast that is animal agriculture and to illuminate it for others.

Truth is a hard apple to catch and a hard apple to throw, I once read. I have had the uncomfortable, guilt-ridden sense of using people—of being a sort of spy, a fly on the wall. I have disliked, by my very presence at animal factories, the appearance of being in collusion with practices I find unconscionable. I have abhorred lying, but am slightly mollified by the fact that the deceptions I engaged in are incredibly minor in comparison to those of agribusiness.

In writing this book, I have felt at times as though I am writing about someone else—someone whom I know, but who is not me. This is because I changed tremendously over the course of my journey.

I went from a Wall Streeter to an investigator. I lived with strangers. I rode trucks and tractors, and snowmobiles and motorbikes. I traveled and hitchhiked. I sought out the truth even at the cost of repelling myself. My journey was profound and my transformation complete.

During my first farm stay, with the Millers, I had not known the difference between hay and straw. During my ensuing stay with the Roberts family, I'd been trying to get Uncle Brick to take me to his egg operation—not realizing that I'd been there the whole time. By the conclusion of my international trek, I was able to recognize animal factories from a distance, and I even spoke like a farmer.

Now it was farmers who asked me my opinion, instead of the other way around. Among the last farms I visited was a beautiful, pastoral, Mennonite organic dairy in Canada. The kind, elderly, white-haired farmer, to whom I spoke for hours (and who recently died), said: "How do you *know* so much? Your thoughts and ideas are not at all trivial. I'm not used to having such conversations with women."

He didn't understand why I beamed as I did. The reason was that I'd realized I'd come full circle. "We shall not cease from exploration," wrote American poet T. S. Eliot, "and the end of all our exploring will be to arrive where we started and know the place for the first time." I'd arrived where I'd started—dairy—and I *knew* the place fully for the first time.

Over the course of my exploration, I also developed a deeper appreciation of animals and a stronger connection to them. Like most people, I'd hardly ever seen cows, pigs, or chickens before. Now, I found that I liked them.

I became more open-minded. I learned that, as with a book, one must not judge a person by his cover. When I opened the books of others' characters, I found in their pages mystery, complexity, sensitivity, and beauty. I came to believe more in humanity. It was people who propelled me on, intentionally or unintentionally.

I also conquered a fear that meant little to others, but that had, for years, seemed as insurmountable to me as embarking on an expedition to Mars. I learned how to drive. If not for my farm investigations, I would never have learned.

In my teens and early twenties, my opinions of myself were like a scattered collage of images formed by others. Today, I see that I must be the only one who paints my own picture. I no longer wish to wrap myself like an ivy plant around the whims of society.

One of the reasons I got along so well with the men of agribusiness is that I, more than most, understand the business mindset. It was in large part my willingness to engage in detailed discussions of costs and efficiencies that led me to fit in as well as I did. I was from Wall Street, for God's sake; I was hard-nosed and capitalistic, rabidly driven to succeed financially.

But now I see the trouble with such a singular mindset. Agriculture is as it is today *because* of such a mindset. I finally recognize that financial success is not the only form of success that exists. A balance between financial, ethical, and emotional considerations is necessary in all realms of life.

"In the years that you spent on this project, you could have made hundreds of thousands of dollars," my father informed me. He was right, yet his words bounced off me. It was then that I saw definitively that I no longer care for expensive clothes or luxurious apartments.

"You have the joy of possession," my mother observed about me when I was a child.

I have finally discarded that joy like an outgrown garment. Previously, my personality traits were like the things in a thrift shop—collected from here and there, belonging to others, compiled together only by the hand of haphazardness. I have cleaned out my shelves and tossed out the odds and ends of my personality, replacing them with traits that I have myself selected.

In my quest to find solutions for animal agriculture, I stumbled also upon solutions for myself. I became a better person. My journey changed me and I am cautiously optimistic that it changes the world around me, too. The first step to internal and external change both is discovery.

During one of my long stretches of writing, it was evening, and I looked out the window. Pockets of apartments were lit yellow like fireflies, people going on with their lives behind the glass. I realized with a start that those people, unreachable to me otherwise, can be reached by this work. My aim is to connect with people who, even if close, are too far away.

My residence in agriculture became much longer than I'd ever intended. Today, it feels bittersweet to exit an industry to whose people and animals I will always feel indebted. I end my journey with a sigh of relief, a note of goodwill, and a handshake of hope.

NOTES

Accuracy has been my aim throughout this work. The chapters arise from more than 2,000 pages of notes and 2,000 photos. The illustrations have been prepared by me using the photos. I have verified facts independently and have also had others do so. All errors in the work are my own.

To protect personal privacy, many names have been changed. In Chapters One to Eight, occurring in Canada, most names have been changed, with the exception of Chapter Six, because the Harleys are supportive of this project. In Chapters Nine to Twelve, occurring in Asia and Mexico, most first names have been retained but supplied without a last name. In Chapters Thirteen to Fifteen, occurring in Belize and the United States, most names have been changed.

Notwithstanding changes in names, all personal characteristics—age, height, background, personality, etc.—have been meticulously retained. Company names have also been retained. All places, conversations, and characters have been rendered precisely.

ACKNOWLEDGMENTS

It is impossible to not feel overwhelmed with gratitude to all who have made this project possible. I lean on too many shoulders to count, and have racked up too many debts to repay.

I would like to thank my twin sister, Sofia, for being my best friend and for supporting me from the first word of this work to the last. She reminded me constantly that what I was doing was important and that I should not give up. I would like to thank my husband, Aamer, for his love and understanding, and for encouraging me to write even at the expense of our time together.

I would like to thank my brother, Salman, for his optimism and wit. I would like to thank my parents for their unconditional love and for endowing me with the education and courage that enabled me to embark upon this endeavor. I would like to thank my aunts, uncles, and cousins for their encouragement from the day I was born.

I would like to thank my agent, Laurie Abkemeier of DeFiore and Company, for advising me through the task of publishing. I would like to thank my editor, Jessica Case of Pegasus Books, for her support. I would also like to thank my interns for their assistance with book promotion. In alphabetic order, Alek Abate, Jess Beaulieu, Rachel Berardinelli, Elizabeth Buff, Raj Ghatore, Janelle Kuehnert, Autumn Ladouceur, Chris Liptrot, Ashley Rawlings,

Ashley Richardson, Ashley Ryan, and Halley Simpson have given generously of their time and considerable skills. I am inspired by their energy and efforts. I would also like to thank Randy Abate for his enthusiastic support and encouragement.

Without farmers, there would be no book. It is their opinions and stories that form the flesh and blood of these pages. Some farmers deserve special gratitude and occupy a special place in my heart. If Brick Roberts had not welcomed me with open arms into his home and community, everything would have ended before it could have begun. Roger Harley spent days explaining the intricacies of agriculture to me and has been an enthusiastic proponent of this project from the start. Agung from Indonesia inspired me to become a better person.

I am also grateful to you, the reader, for staying with me to the end. I hope you have found in this project what you were looking for, just as I have.

SOURCES

Listed below, by chapter, are the approximately six hundred sources used to supplement the original investigative research of the book.

Chapter One: Organic Dairy Cows

"All differences in one table by EU regulation." Organic Rules and Certification. Accessed February 2015. http://organicrules.org/custom/differences.php?id=2abg.

"Artificial Insemination: What About the Other Animals?" Freakonomics. November 22, 2011. Accessed February 2015. http://freakonomics.com/2011/11/22/artificial-insemination-what-about-the-other-animals/.

Behar, Harriet. "An Overview of the Access to Pasture Rule on Organic Dairy Farms." Extension. March 18, 2014. Accessed February 2015. https://www.extension.org/pages/28875/an-overview-of-the-access-to-pasture-rule-on-organic-dairy-farms.

"Canada's organic market now worth $3.7 billion—Growth driven by broad-scale support of organic foods." *CNW.* April 11, 2013. Accessed February 2015. http://www.newswire.ca/en/story/1144253/canada-s-organic-market-now-worth-3-7-billion-growth-driven-by-broad-scale-support-of-organic-foods.

Chant, Ian. "Unnatural Turkeys: A New Marketplace Podcast." Freakonomics. Accessed February 2015. http://freakonomics.com/2011/11/17/unnatural-turkeys-a-new-marketplace-podcast/.

Ciaq. Accessed February 2015. http://www.ciaq.com/homepage.html.

"Commission Regulation (EC) No 889/2008." European Union. September 5, 2008. Accessed February 2015. http://eurlex.europa.eu/LexUriServ/LexUriServ.do?uri=OJ:L:2008:250:0001:0084:EN:PDF.

"Consumer Victory: USDA Proposes Mandatory Pasture and Feed Requirements for Organic Dairy Farms." Organic Consumers Association. October 23, 2008. Accessed February 2015. https://www.organicconsumers.org/news/consumer-victory-usda-proposes-mandatory-pasture-and-feed-requirements-organic-dairy-farms.

"Council Directive 97/2/EC." European Food Safety Authority. January 20, 1997. Accessed February 2015. http://www.efsa.europa.eu/en/efsajournal/doc/366.pdf.

"Council Regulation (EC) No 834/2007." Council of the European Union. June 28, 2007. Accessed February 2015. http://faolex.fao.org/docs/pdf/eur72928.pdf.

"Cow Trainer Installation and Maintenance." Ontario Ministry of Agriculture, Food and Rural Affairs. July 16, 2005. Accessed February 2015. http://www.mrec.org/pubs/trainer%20portrait%2016july05.pdf.

"Dairy Outlook Overview." Iowa State University Extension Farm. January 2013. Accessed February 2015. http://www.extension.iastate.edu/dairyteam/sites/www.extension.iastate.edu/files/dairyteam/2013.January_outline-ks.pdf.

"December 2014 Proof List." EastGen. Accessed February 2015. http://www.eastgen.ca/i?lang=en&view=list.

"Electronic Code of Federal Regulations: Livestock living conditions." U.S. Government Publishing Office. February 17, 2010. Accessed February 2015. http://www.ecfr.gov/cgi-bin/text-idx?SID=a596d3375263df703a8484a138235e64&node=se7.3.205_1239&rgn=div8.

"Electronic Code of Federal Regulations: Pasture practice standard." U.S. Government Publishing Office. February 17, 2010.

"Farm numbers dip, but organic farms up: Statistics Canada." The Western Producer. June 19, 2012. Accessed February 2015. http://www.producer.com/2012/06/farm-numbers-dip-but-organic-farms-up-statistics-canada/.

Fusaro, Dave. "Organic Sales in 2013 Grew Nearly 12 Percent." *Food Processing*, May 15, 2014. Accessed February 2015. http://www.foodprocessing.com/industrynews/2014/organic-sales-2013/.

"General Principle and Management Standards." Public Works and Government Services Canada. Last modified November 19, 2014. http://www.tpsgc-pwgsc.gc.ca/ongc-cgsb/programme-program/normes-standards/internet/bio-org/principes-principles-eng.html.

Grandin, Temple, and Catherine Johnson. *Animals in Translation*. New York: Scribner, 2004.

"History of the Organic Movement." The Organics Institute. Accessed February 2015. http://theorganicsinstitute.com/organic/history-of-the-organic-movement/.

"Holstein Breed History." Holstein Association USA. Accessed February 2015. http://www.holsteinusa.com/holstein_breed/holstein101.html?tab=1#TabbedPanels1.

Hultgren, Jan. "Cattle Welfare Aspects of Animal Hygiene." International Society for Animal Hygiene. Accessed February 2015. http://www.isah-soc.org/documents/mainspeakers/6%20HultgrenSweden.doc.

Huxley, Aldous. *Brave New World*. Reprint, New York: HarperCollins, 2006.

In Organic We Trust. Film. Directed by Kip Pastor. 2012. Los Angeles, CA: Pasture Pictures. 2013. DVD.

Liebenson, Donald. "For Americans, Healthy Lifestyle Includes Organic Food." Millionaire Corner. August 8, 2014. Accessed February 2015. http://millionairecorner.com/Content_Free/organic-foods-healthy-lifestyle.aspx.

Lush, Jay. *Animal Breeding Plans*. Reprint, Ulan Press, 2012.

"National Organic Standards Board." United States Department of Agriculture—Agricultural Marketing Service. Last modified September 16, 2013. http://www.ams.usda.gov/AMSv1.0/NOSB.

"Organic Agriculture." United States Department of Agriculture. Last modified January 9, 2015. http://www.usda.gov/wps/portal/usda/usdahome?contentidonly=true&contentid=organic-agriculture.html.

"Organic diets becoming more mainstream." *Examiner.com*, August 7, 2014. Accessed February 2015. http://www.examiner.com/article/organic-diets-becoming-more-mainstream.

"Organic Market Overview." United States Department of Agriculture, Economic Research Service. Last modified April 7, 2014. http://www.ers.usda.gov/topics/natural-resources-environment/organic-agriculture/organic-market-overview.aspx.
"Organic Philosophy Report." The Kerr Center for Sustainable Agriculture. August 15, 2014. Accessed February 2015. http://kerrcenter.com/publication/brief-overview-history-philosophy-organic-agriculture/organic-philosophy-report/.
"Organic Production—Canadian Industry." Government of Canada. Last modified October 7, 2014. http://www.agr.gc.ca/eng/industry-markets-and-trade/statistics-and-market-information/by-product-sector/organic-products/organic-production-canadian-industry/?id=1183748510661.
"Organic Rules and Certification." International Centre for Research in Organic Food Systems. Accessed February 2015. http://organicrules.org/custom/differences.php?id=2abi.
Orwell, George. *1984*. New York: Signet Classics, 1950.
"Overview." United States Department of Agriculture, Economic Research Service. Last modified September 27, 2013. http://www.ers.usda.gov/data-products/organic-production/documentation.aspx.
Semex. Accessed February 2015. http://www.semexusa.com/.
"Technical and Learning Resources." Joint FAO/IAEA Programme. Accessed February 2015. http://www-naweb.iaea.org/nafa/aph/resources/technology-ai.html.
"Trout Lifecycle." The Wild Trout Trust. Accessed February 2015. http://www.wildtrout.org/content/trout-lifecycle.
"Turkey: Keeping Turkey as Pets." Dog Breed Info. Accessed February 2015. http://www.dogbreedinfo.com/pets/turkey.htm.
"USDA Issues Final Rule on Organic Access to Pasture." United States Department of Agriculture—Agricultural Marketing Service. Last modified February 12, 2010.
"Using Artificial Insemination in Swine Production: Detecting and Synchronizing Estrus and Using Proper Insemination Technique." Virginia Cooperative Extension. May 1, 2009. Accessed February 2015. http://pubs.ext.vt.edu/414/414-038/414-038.html.

Chapter Two: Egg Agony

Andrews, James. "European Union Bans Battery Cages for Egg-Laying Hens." *Food Safety News*, January 19, 2012. Accessed February 2015. http://www.foodsafetynews.com/2012/01/european-union-bans-battery-cages-for-egg-laying-hens/.
Arndt, Milton. *Battery Brooding*. Reprint, Whitefish, MT: Kessinger Publishing, LLC, 2010.
Balance, Desmond. "Producing Omega-3 Enriched Eggs." Flax Council. Accessed February 2015. http://www.flaxcouncil.ca/english/pdf/eggsbalance.pdf.
Barnes, John, and Aziz Tahseen. "Harmful effects of ammonia on birds." World Poultry. October 25, 2010. Accessed February 2015. http://www.worldpoultry.net/Breeders/Health/2010/10/Harmful-effects-of-ammonia-on-birds-WP008071W/.
"Caged Hens Free from Debeaking." The Poultry Site. July 18, 2013. Accessed February 2015. http://www.thepoultrysite.com/poultrynews/29517/caged-hens-free-from-debeaking.
Cheng, H. "Morphopathological changes and pain in beak trimmed laying hens." *World's Poultry Science Journal* 62, no. 1 (2006): 41-52. Accessed February 2015. doi: 10.1079/WPS200583.
"Council Directive laying down minimum standards for the protection of laying hens." EUR-LEX. July 19, 1999. Accessed February 2015. http://eur-lex.europa.eu/legal-content/EN/TXT/?uri=CELEX:31999L0074.

"Debeaking." United Poultry Concerns, Inc. Accessed February 2015. http://www.upc-online.org/merchandise/debeak_factsheet.html.

Duncan, Ian. "Behavior and Behavioral Needs." *Poultry Science* 77, no. 12 (1999): 1766-1772. Accessed February 2015. doi: 10.1093/ps/77.12.1766.

Encyclopaedia of Occupational Health and Safety, Fourth Edition, Edited by Jeanne Mager Stellman. Geneva: International Labor Org, 1998.

Grandin, Temple. "Corporations Can be Agents of Great Improvements In Animal Welfare and Food Safety and the Need for Minimum Decent Standards." National Institute of Animal Agriculture. April 4, 2001. Accessed February 2015. http://www.grandin.com/welfare/corporation.agents.html.

Harrison, Ruth. *Animal Machines*. London: Vincent Stuart Publishers, Ltd., 1964.

Hester, P.Y., and M. Shea-Moore. "Beak trimming egg-laying strains of chickens." *World's Poultry Science Journal* 59, no. 4 (2003): 458-474. Accessed February 2015. doi: 10.1079/WPS20030029.

"History of Egg Production." American Egg Board. Accessed February 2015. http://www.aeb.org/farmers-and-marketers/history-of-egg-production.

Kowol, Adam. "The theory of cognitive dissonance." Adam Kowol. 2008. Accessed February 2015. http://works.adamkowol.info/Festinger.pdf.

McLeod, Saul. "Cognitive Dissonance." Simply Psychology. Last modified 2014. Accessed February 2015. http://www.simplypsychology.org/cognitive-dissonance.html.

Mench, Joy. "Why it is important to Understand Animal Behavior." *ILAR Journal* 39, no. 1 (1998): 20-26.

"More Humane Egg Production." *New York Times*, February 14, 2012. Accessed February 2015. http://www.nytimes.com/2012/02/15/opinion/more-humane-egg-production.html?_r=0.

"Opinion of the Scientific Panel on Animal Health and Welfare (AHAW) on a request from the Commission related to the welfare aspects of various systems of keeping laying hens." European Food Safety Authority. March 7, 2005. Accessed February 2015. http://www.efsa.europa.eu/en/efsajournal/pub/197.htm.

Orwell, George. *1984*. New York: Signet Classics, 1950.

Pickett, Heather. "Controlling Feather Pecking & Cannibalism in Laying Hens Without Beak Trimming." Compassion in World Farming. July 2008. Accessed February 2015. https://www.ciwf.org.uk/includes/documents/cm_docs/2008/c/controlling_feather_pecking_and_cannibalism_in_laying_hens.pdf.

"Prolapsed Oviduct, Prolapsed Vent, or Blowout." PoultryHelp.com. Last modified March 31, 2005. http://www.poultryhelp.com/prolapse.html.

Rodenburg, T.B., F.A.M. Tuyttens, K. de Reu, L. Herman, J. Zoons, and B. Sonck, "Welfare assessment of laying hens in furnished cages and non-cage systems: an on-farm comparison." *Animal Welfare* 17, no. 4 (2008): 363-373.

Shields, Sara, and Ian Duncan. "A Comparison of the Welfare of Hens in Battery Cages and Alternative Systems." The Humane Society of the United States.

Steinfeld, Henning, et al. "Livestock's long shadow: environmental issues and options." Rome: Food and Agriculture Organization of the United Nations, 2006. http://www.europarl.europa.eu/climatechange/doc/FAO%20report%20executive%20summary.pdf.

Tauson, Ragnar. "Mortality in Laying Hens Caused by Differences in Cage Design." *Acta Agric. Scand.* 35 (1985): 165-174.

"The Welfare of Animals in the Egg Industry." The Humane Society of the United States.
"Welfare Implications of Beak Trimming." American Veterinary Medical Association—Animal Welfare Division. February 7, 2010.
"Welfare implications of changes in production systems for laying hens." European Commission. March 28, 2006. Accessed February 2015. http://ec.europa.eu/food/animal/welfare/farm/laywel_final_report_en.pdf.
Williams, Erin, and Margo DeMello. *Why Animals Matter.* Amherst, New York: Prometheus Books, 2007.

Chapter Three: Porkmaking by Machinery

"Agribusiness and Agri-Food." Farm Credit Canada. Accessed February 2015. https://www.fcc-fac.ca/en/we-finance/agribusiness-and-agri-food.html.
"Alternatives for Pig Castration." Alternative Pig. Accessed February 2015. http://www.alternativepig.eu/home.html.
Brodhagen, Amanda. "Pig Barn Explosion, Leaves Two Farm Workers Seriously Injured." Farms.com. October 3, 2012. Accessed February 2015. http://www.farms.com/ag-industry-news/pig-barn-explosion-leaves-two-farm-workers-seriously-injured-770.aspx.
Conestoga Meat Packers. Accessed February 2015. http://conestogameats.com/.
Dingman, Kali. "Exploding hog barns beckon U researchers." *Minnesota Daily,* February 7, 2012. Accessed February 2015. http://www.mndaily.com/2012/02/07/exploding-hog-barns-beckon-u-researchers.
Drung, Arnold. "Integration from Breeding to Feeding to Eating—A Producer Owned Cooperative." *Advances in Pork Production* 17 (2006): 87. Accessed February 2015. http://www.prairieswine.com/pdf/2469.pdf.
Duncan, Ian. "Behavior and Behavioral Needs." *Poultry Science* 77, no. 12 (1999): 1766-1772. Accessed February 2015. doi: 10.1093/ps/77.12.1766.
"European Pig Castration Campaign." Eurogroup for Animals. Accessed February 2015. http://eurogroupforanimals.org/get-involved/european-pig-castration-campaign/.
Grandin, Temple. "Corporations Can be Agents of Great Improvements In Animal Welfare and Food Safety and the Need for Minimum Decent Standards." National Institute of Animal Agriculture. April 4, 2001. Accessed February 2015. http://www.grandin.com/welfare/corporation.agents.html.
Grandin, Temple, and Catherine Johnson. *Animals in Translation*. New York: Scribner, 2004.
Ketchem, Ron, and Mark Rix. "How to Manage Recycling Females to Improve Farrowing Rate, Total Pigs Born." National Hog Farmer. May 7, 2012.
"Lameness." The Pig Site. Accessed February 2015. http://www.thepigsite.com/diseaseinfo/57/lameness.
Orwell, George. *Animal Farm*. Reprint, New York: Signet Classics, 1996.
"Percent of consumer expenditures spent on food, alcoholic beverages, and tobacco that were consumed at home, by selected countries, 2012; Per capita food expenditures; and Percent of consumer expenditures spent on food, alcoholic beverages, and tobacco that were consumed at home, by selected countries, 2012." United States Department of Agriculture—Economic Research Service. Last modified February 18, 2015.
"Piglet Losses." University of Illinois Extension. Accessed February 2015. http://livestocktrail.illinois.edu/swinerepronet/paperDisplay.cfm?ContentID=6266.

"Prescription for Trouble: Using Antibiotics to Fatten Livestock." Union of Concerned Scientists. Accessed February 2015. http://www.ucsusa.org/food_and_agriculture/our-failing-food-system/industrial-agriculture/prescription-for-trouble.html#.VIsaDjHF-So.

"Research Reviews Top Reasons Producers Cull Sows." National Hog Farmer. March 23, 2012. Accessed February 2015. http://nationalhogfarmer.com/reproduction/research-reviews-top-reasons-producers-cull-sows.

"The risks of poor welfare in intensive calf farming systems." *European Food Safety Authority* 366 (2006): 1-36. Accessed February 2015. http://www.efsa.europa.eu/en/efsajournal/doc/366.pdf.

Schmitt, Kristen. "Deer Farming: The Next Adventure in Agriculture." Modern Farmer. February 19, 2014. Accessed February 2015. http://modernfarmer.com/2014/02/deer-farming-next-adventure-agriculture/.

Scientific Veterinary Committee. "The Welfare of Intensively Kept Pigs." European Commission. September 30, 1997. Accessed February 2015. http://ec.europa.eu/food/fs/sc/oldcomm4/out17_en.pdf.

Sinclair, Upton. *The Jungle*. Reprint, Mineola, NY: Dover Publications, 2001.

Tolstoy, Leo. *The First Step: An Essay on the Morals of Diet*. Albert Broadbent, 1900.

Turner, Jacky. *Animal Breeding, Welfare, and Society*. New York: Routledge, 2010.

Vansickle, Joe. "Pork Board Addresses Penicillin Residues in Sows." National Hog Farmer. August 19, 2013. Accessed February 2015. http://nationalhogfarmer.com/health/pork-board-addresses-penicillin-residues-sows.

Vansickle, Joe. "Watch Out for Foaming Manure." National Hog Farmer. March 15, 2010. Accessed February 2015. http://nationalhogfarmer.com/mag/farming_watch_foaming_manure.

"Welfare on the Farm—Ongoing Initiatives." European Commission. Accessed February 2015. http://ec.europa.eu/food/animal/welfare/farm/initiatives_en.htm.

"What is penicillin?" Medical News Today. Last modified September 26, 2014. Accessed February 2015. http://www.medicalnewstoday.com/articles/216798.php.

White, E.B. *Charlotte's Web*. Reprint, New York: HarperCollins, 2012.

White, E.B. "Death of a Pig." *The Atlantic* 181, no. 1 (1948): 28-33. Accessed February 2015.

Willyard, Cassandra. "Mysterious Exploding Foam is Bursting Barns." Smithsonian.com. March 14, 2012. Accessed February 2015. http://blogs.smithsonianmag.com/science/2012/03/mysterious-exploding-foam-is-bursting-barns/.

Wilson, Edward O. *On Human Nature*. Revised, Cambridge, MA: Harvard University Press, 2004.

Wright, Andy. "Pigheaded: How Smart are Swine?" Modern Farmer. March 10, 2014. Accessed February 2015. http://modernfarmer.com/2014/03/pigheaded-smart-swine/.

"You Can Save More Pigs from Those Large Litters." National Hog Farmer. May 9, 2011. Accessed February 2015. http://nationalhogfarmer.com/genetics-reproduction/you-can-save-more-pigs-from-large-litters-0509/.

Zaitchik, Alexander. "Big Ag's Big Lie: Factory Farms, your Health and the new politics of antibiotics." Salon. January 12, 2014. Accessed February 2015. http://www.salon.com/2014/01/12/big_ags_big_lie_factory_farms_your_health_and_the_new_politics_of_antibiotics/.

Chapter Four: Trilling Turkeys

"Meat and Poultry Labeling Terms." United States Department of Agriculture—Food Safety and Inspection Service. Last modified April 2011. Accessed February 2015. http://www.fsis.usda.gov/wps/wcm/connect/e2853601-3edb-45d3-90dc-1bef17b7f277/Meat_and_Poultry_Labeling_Terms.pdf?MOD=AJPERES.

"Meat and Poultry Labeling Terms." United States Department of Agriculture—Food Safety and Inspection Service. Last modified October 24, 2014. Accessed February 2015. http://www.fsis.usda.gov/wps/portal/fsis/topics/food-safety-education/get-answers/food-safety-fact-sheets/food-labeling/meat-and-poultry-labeling-terms/meat-and-poultry-labeling-terms.

Merica, Dan. "Where pardoned turkeys go to die." *CNN*, November 27, 2013. Accessed February 2015. http://www.cnn.com/2013/11/27/politics/pardoned-turkeys/.

Mroz, Emilia, and Aneta Orlowska. "Quality of heavy-type turkey poults as related to the age of layers in the first laying season." *Animal Science Papers and Reports* 27, no. 3 (2009): 207-215. Accessed February 2015. http://www.ighz.edu.pl/files/objects/7511/66/strona207-216.pdf.

"National Organic Program." United States Department of Agriculture—Agricultural Marketing Service. Last modified October 17, 2012. Accessed February 2015. http://www.ams.usda.gov/AMSv1.0/NOPConsumers.

National Turkey Federation. Accessed February 2015. http://www.eatturkey.com/.

Poultry Science Association. "Reducing femur fractures in turkeys." Ag Annex. November 2012. Accessed February 2015. http://www.agannex.com/production/reducing-femur-fractures-in-turkeys.

"President Obama Pardons White House Turkey." *YouTube*. The White House. 2011. https://www.youtube.com/watch?v=9J9PV7IdrdU.

Sinclair, Upton. *The Jungle*. Reprint, Mineola, NY: Dover Publications, 2001.

"Toe Trimming of Turkeys or Chickens in Small and Backyard Poultry Flocks." Extension. November 12, 2012. Accessed February 2015. http://www.extension.org/pages/66246/toe-trimming-of-turkeys-or-chickens-in-small-and-backyard-poultry-flocks.

Chapter Five: Crank Chicken

Bottemiller, Helena. "Q&A With Temple Grandin Part II." *Food Safety News*, June 23, 2010. Accessed February 2015. http://www.foodsafetynews.com/2010/06/qa-with-temple-grandin-part-ii/#.Ul7VVFCkeSo.

Clancy, Heather. "These days, Old MacDonald has a farm and a smartphone." ZDNet. September 14, 2012. Accessed February 2015. http://www.smartplanet.com/blog/business-brains/these-days-old-macdonald-has-a-farm-and-a-smartphone/26037.

Commercial Chicken Meat and Egg Production, Edited by Donald Bell and William Weaver. Boston, MA: Kluwer Academic Publishers, 2002.

Darling Ingredients Inc. Accessed February 2015. http://www.darlingii.com/.

"Darling Ingredients Inc. To Acquire The Rothsay Rendering Business From Maple Leaf Foods Inc." Darling International Inc. August 23, 2013. Accessed February 2015. http://ir.darlingii.com/profiles/investor/NewsPrint.asp?b=1640&ID=64711&m=rl&pop=1&G=388&Nav=0.

DeYoung, Jeff. "Smartphone puts control of farm at fingertips." Iowa Farmer Today. July 3, 2013. Accessed February 2015. http://www.iowafarmertoday.com/news/

crop/smartphone-puts-control-of-farm-at-fingertips/article_0af58e0e-e34b-11e2-9722-001a4bcf887a.html.

Ensminger M.E., J.E. Oldfield, and W.W. Heinemann. *Feeds and Nutrition Digest.* Clovis, CA: Ensminger Publishing Co. 1990.

Firman, Jeffre. "Rendered Products in Poultry Nutrition." National Renderers Association. 2006. Accessed February 2015. http://assets.nationalrenderers.org/essential_rendering_poultry.pdf.

Grand River Foods. Accessed February 2015. http://www.grandriverfoods.com.

Grandin, Temple, and Catherine Johnson. *Animals in Translation*. New York: Scribner, 2004.

Havenstein, G.B., P.R. Ferket, and M.A. Qureshi. "Growth, livability, and feed conversion of 1957 versus 2001 broilers when fed representative 1957 and 2001 broiler diets." *Poultry Science* 82, no. 10 (2003): 1500-1508. Accessed February 2015. doi: 10.1093/ps/82.10.1500.

Kawar, Mark. "Company gets picky about what smart chickens eat." *Omaha World-Herald*, March 5, 2003.

"Latest Annual Reports." Maple Leaf Foods. Accessed February 2015. http://investor.mapleleaf.ca/phoenix.zhtml?c=88490&p=irol-reportsannual.

Maple Leaf Foods. Accessed February 2015. http://www.mapleleaffoods.com/.

Marti, Daniel, Rachel Johnson, and Kenneth Mathews, Jr. "Where's the (Not) Meat?" United States Department of Agriculture—Economic Research Service. November 2011. Accessed February 2015. http://www.ers.usda.gov/media/147867/ldpm20901.pdf.

McGlone, John, and Wilson Pond. *Pig Production: Biological Principles and Applications*. Cengage Learning, 2002.

"Number of farmers continue steep decline." *The Leader-Post*, December 3, 2008. Accessed February 2015. http://www.canada.com/reginaleaderpost/news/story.html?id=5decacf8-b6b4-4212-b5b4-874834f08b67.

Pollan, Michael. *The Omnivore's Dilemma*. New York: Penguin Books, 2006.

"Poultry Nutrition." National Renderers Association. Accessed February 2015. http://www.nationalrenderers.org/nutrition/poultry/.

Ross. Accessed February 2015. http://en.aviagen.com/ross/.

Rothsay. Accessed February 2015. http://www.rothsay.ca/.

Schlosser, Eric. *Fast Food Nation*. New York: Houghton Mifflin, 2001.

"U.S. Broiler Performance." National Chicken Council. Last modified January 9, 2015. Accessed February 2015. http://www.nationalchickencouncil.org/about-the-industry/statistics/u-s-broiler-performance/.

Villalba, J.J., F.D. Provenza, and X. Manteca. "Links between ruminants' food preference and their welfare." *Animal* 4, no. 7 (2010): 1240-1247. Accessed February 2015. doi: 10.1017/S1751731110000467.

Winfield, R.G., and J.A. Monroe. "Stray Voltage." Canada Plan Service. Accessed February 2015. http://www.cps.gov.on.ca/english/plans/E9000/9611/M-9611L.pdf.

Woodill, Gary, and Chad Udell. "mAgriculture: The Application of Mobile Computing to the Business of Farming; mAgriculture: Knowing What you Eat—Using Mobile Apps to Track Food from Farm to Table; and mAgriculture: Future Uses of Mobile Technologies in Farming, Fishing, Forestry." Float Mobile Learning. 2012. Accessed February 2015. http://floatlearning.com/download-mobile-agriculture-environmental-scans/?submissionGuid=84ff7d45-b3ee-44ee-bc48-77be9f9fdf65.

Chapter Six: Animal Heaven

Avery, Alex. *The Truth About Organic Foods*. St. Louis: Henderson Communications L.L.C., 2006.

Cox, Billy, and Hakim Fobia. "USDA Establishes Naturally Raised Marketing Claim Standard." United States Department of Agriculture Agricultural Marketing Service. January 16, 2009.

"Fact Sheet: Organic Products in Canada." Canadian Food Inspection Agency. Last modified February 17, 2014. http://www.inspection.gc.ca/food/organic-products/fact-sheet/eng/1389651477171/1389651725636.

"Food expenditures by families and individuals as a share of disposable personal income." U.S. Department of Agriculture. 2013.

Gandhi, Mahatma K. *Harijan* (1940): 281.

Gandhi, Mahatma K. *Young India* (1925): 8.

Harley Farms. Accessed February 2015. http://harleyfarms.ca/.

In Organic We Trust. Film. Directed by Kip Pastor. 2012. Los Angeles, CA: Pasture Pictures. 2013. DVD.

"Method of Production Claims." Canadian Food Inspection Agency. Last modified May 26, 2014. http://www.inspection.gc.ca/food/labelling/food-labelling-for-industry/method-of-production-claims/eng/1389379565794/1389380926083.

"Per capita food expenditures." U.S. Department of Agriculture. 2013.

"Percent of consumer expenditures spent on food, alcoholic beverages, and tobacco that were consumed at home, by selected countries, 2012." U.S. Department of Agriculture. 2013.

"Pet Industry Market Size & Ownership Statistics." American Pet Products Association. Accessed February 2015. http://www.americanpetproducts.org/press_industrytrends.asp.

Plain, Ron, John Lawrence, and Glenn Grimes. "The Structure of the U.S. Pork Industry." U.S. Pork Center of Excellence. 2006. Accessed February 2015.

Rowe Farms. Accessed February 2015. http://www.rowefarms.ca/.

Chapter Seven: Red, White, and Bob Veal

"All About Veal." Ontario Veal. Accessed February 2015. http://ontarioveal.on.ca/all-about-veal/.

American Veal Association. Accessed February 2015. http://www.americanveal.com/.

American Veterinary Medical Association. "Literature Review on the Welfare Implications of the Veal Calf Husbandry." October 13, 2008. Accessed February 2015.

Bain, Jennifer. "The Real Deal About Veal." *The Star*, April 4, 2007. Accessed February 2015. http://www.thestar.com/life/food_wine/recipes/2007/04/04/the_real_deal_about_veal.html.

Black, Jane. "The Kinder Side of Veal." *The Washington Post*, October 28, 2009. Accessed February 2015. http://www.washingtonpost.com/wp-dyn/content/article/2009/10/27/AR2009102700563.html.

Clark Eggs Farms Ltd. Accessed February 2015. http://www.clarkeggfarms.com/.

Cook, Angela, Richard J. Reid-Smith, Rebecca J. Irwin, Scott A. McEwen, Virginia Young, and Carl Ribble. "Antimicrobial resistance in *campylobacter*, *salmonella*, and *Escherichia coli* isolated from retail grain-fed veal meat from southern Ontario, Canada." *Journal of Food Protection* 74 (2011): 1245-1251.

European Food Safety Authority. "Scientific Opinion on the risks of poor welfare in intensive calf farming systems. An update of the Scientific Veterinary Committee Report on the Welfare of Calves." *The EFSA Journal* 366 (2006): 1-36.

"Facts & Figures About Canadian Veal." Farm & Food Care. Accessed February 2015. http://www.farmfoodcare.org/.

"Factsheet: *Salmonella* Dublin." Ontario Veal. Accessed February 2015. http://ontarioveal.on.ca/factsheet-salmonella-dublin/.

Farm & Food Care. Accessed February 2015. http://www.farmfoodcare.org/.

Father of the Bride. Directed by Charles Shyer. 1991. Burbank, CA: Touchstone Pictures.

Greter, Angela, and Léna Levison. "Calf in a Box: Individual Confinement Housing Used in Veal Production." BC SPCA Farm Animal Welfare News. June 2012. Accessed February 2015. http://www.spca.bc.ca/assets/documents/welfare/farm/farmsense/june-2012-feature-story.pdf.

Harrison, Ruth. *Animal Machines*. London: Vincent Stuart Publishers, Ltd., 1964.

Hickman, Martin. "The Ethics of Eating: The Appeal of Veal." *The Independent*, September 2, 2006. Accessed February 2015. http://web.archive.org/web/20071012074555/http:/www.independent.co.uk/living/food_and_drink/news/article1269468.ece.

"History of Veal." London Veal. Accessed February 2015. http://londonveal.com/?page_id=89.

"Humane Methods of Slaughter Act: Actions are needed to strengthen enforcement." US GAO. Accessed February 2015. http://www.gao.gov/products/GAO-10-203.

"Issues Centre." Ontario Farm Animal Council. Accessed February 2015. http://archive.is/ySgaB.

LeTrent, Sarah. "Targeting consumers' beef with veal." *CNN*, August 6, 2013. Accessed February 2015. http:/eatocracy.cnn.com/2013/08/06/targeting-consumers-beef-with-veal/.

Neindre, P. Le. "Evaluating housing systems for veal calves." *Journal of Animal Science* 71 (1993): 1345-1354.

Ngapo, Tania M., and Claude Gariépy. "Factors affecting the meat quality of veal." *Journal of the Science of Food and Agriculture* 86 (2006): 1412-1431.

Ontario Veal. Accessed February 2015. http://ontarioveal.on.ca/.

Ontario Veal Appeal. Accessed February 2015. http://ontariovealappeal.ca/.

Picard, Ken. "More legal woes for Bushway slaughterhouse." Seven Days. March 23, 2011. Accessed February 2015. http://www.sevendaysvt.com/vermont/more-legal-woes-for-bushwayslaughterhouse/Content?oid=2142875.

"Questions and Answers." Veal Farm. Accessed February 2015. http://www.vealfarm.com/questions-and-answers/.

"The Real Deal About Veal!" Ontario Veal Association. http://ontarioveal.on.ca/all-about-veal/the-real-deal-about-veal/.

Solotaroff, Paul. "In the Belly of the Beast." Rolling Stone. December 10, 2013. Accessed February 2015. http://www.rollingstone.com/feature/belly-beast-meat-factory-farms-animal-activists.

Specht, Joshua. "A failure to prohibit: New York City's underground bob veal trade." *The Journal of the Gilded Age and Progressive Era* 12 (2013): 475-501.

Stacey, Caroline. "Is veal cruel?" *British Broadcasting Company*, May 29, 2007. Accessed February 2015. http://web.archive.org/web/20070529064739/http://www.bbc.co.uk/food/food_matters/veal.shtml.

"Undercover Investigation: Abused Calves at Vermont Slaughter Plant." *YouTube* video. 4:11, posted by "VIN Vid." November 3, 2009. Accessed February 2015. https://www.youtube.com/watch?v=iiBP7umdPTg.

"Undercover video shows 'abuse' of Quebec veal calves." *CBCNews Montreal*, April 21, 2014. Accessed February 2015. http://www.cbc.ca/news/canada/montreal/undercover-video-shows-abuse-of-quebec-veal-calves-1.2616716.

Vallely, Paul. "For What Cause Did Jill Phipps Die?" *The Independent*, February 3, 1995. Accessed February 2015. http://www.independent.co.uk/news/uk/for-what-cause-did-jill-phipps-die-1571300.html.

"Veal Farms." Virtual Farm Tours. Accessed February 2015. http://www.virtualfarmtours.ca/en/vealFarms/index.html.

"Veal farmers move calves to group pens." Beef Magazine. May 9, 2012. Accessed February 2015. http://beefmagazine.com/sectors/veal-farmers-move-calves-group-pens.

"Veal Farming." Veal Farm. Accessed February 2015. http://www.vealfarm.com/veal-farming/.

"Veal from farm to table." United States Department of Agriculture Food Safety and Inspection Service Food Safety Information. June 2011. Accessed February 2015. http://www.fsis.usda.gov/wps/wcm/connect/c1c3ed6a-c1e5-4ad0-ba6c-d53d71d741c6/Veal_from_Farm_to_Table.pdf?MOD=AJPERES.

"The Veal Information Gateway." Veal. Accessed February 2015. http://www.veal.ca/.

Veal: Supply and disappearance (carcass weight, million pounds) and per capita disappearance (pounds)." U.S. Department of Agriculture. 2014.

What Do Cows Do? TickTock Books Ltd., 2008.

Chapter Eight: Slaughter Horror

"Annex A—species-specific stunning guidelines—red meat species." Canadian Food Inspection Agency. Last modified July 20, 2013. http://www.inspection.gc.ca/food/meat-and-poultry-products/manual-of-procedures/chapter-12/annex-a/eng/1374358238209/1374358242162?chap=3.

Barclay, Christopher. *Religious Slaughter.* House of Commons Library, 2011.

Becker, Geoffrey S. "CRS Report for Congress: Nonambulatory Livestock and the Humane Methods of Slaughter Act." Congressional Research Service. March 24, 2009. Accessed February 2015.

Becker, Geoffrey S. "CRS Report for Congress: USDA Meat Inspection and the Humane Methods of Slaughter Act." Congressional Research Service. February 26, 2008. Accessed February 2015.

Canadian Minister of Justice. "Meat Inspection Regulations, 1990." Ottawa, Canada: 1990.

"Cattlemen's college: Cattle feed efficiency." Agriculture Proud. Last modified February 6, 2012. http://agricultureproud.com/2012/02/06/cattlemens-college-cattle-feed-efficiency/.

"Chapter 12: Food animal humane handling and slaughter—animal welfare requirements." Canadian Food Inspection Agency. Last modified October 1, 2014. http://www.inspection.gc.ca/food/meat-and-poultry-products/manual-of-procedures/chapter-12/animal-welfare-requirements/eng/1392144659190/1392144660111?chap=0.

Dillard, Jennifer. "A slaughterhouse nightmare: Psychological harm suffered by slaughterhouse employees and the possibility of redress through legal reform." *Georgetown Journal on Poverty Law & Policy* 15 (2008): 391-408.

Fitzgerald, Amy J. "A social history of the slaughterhouse: From inception to contemporary implications." *Human Ecology Review* 17 (2010): 58-69.

Fitzgerald, Amy J., Linda Kalof, and Thomas Dietz. "Slaughterhouses and increased crime rates: An empirical analysis of the spillover from 'The Jungle' into the surrounding community." *Organization Environment* 22 (2009): 158-184.

Food Safety and Inspection Service. "Availability of FSIS Compliance Guide for a Systematic Approach to the Humane Handling of Livestock." Federal Register. October 29, 2013.

"How to determine insensibility in cattle, pigs, and sheep in slaughter plants." Dr. Temple Grandin's Web Page. Last modified July 2012. http://www.grandin.com/humane/insensibility.html.

"The HSUS mourns death of Dr. Dean Wyatt." The Humane Society of the United States. Last modified November 9, 2010. http://www.humanesociety.org/news/press_releases/2010/11/dr_dean_wyatt_mourned_110910.html.

Humane Methods of Slaughter Act. U.S. Code 7 (2011), § 1901-1907.

"Humane Methods of Slaughter Act: Actions are needed to strengthen enforcement." US GAO. February 2010. http://www.gao.gov/products/GAO-10-203.

Humane Methods of Slaughter Act of 1978. Pub. L. No. 95-445, 95th Cong., 1st Sess. (October 10, 1978), GPO.

"Humane Methods of Slaughter Act: USDA has addressed some problems but still faces enforcement challenges." US GAO. January 2004. http://www.gao.gov/new.items/d04247.pdf.

Lappé, Frances Moore. *Diet for a Small Planet*. New York: Ballantine Books, 1971.

Marti, Daniel L., Rachel J. Johnson, and Kenneth H. Mathews. "Where's the (Not) Meat? Byproducts From Beef and Pork Production." United States Department of Agriculture, 2011. http://www.ers.usda.gov/publications/ldpm-livestock,-dairy,-and-poultry-outlook/ldpm209-01.aspx.

"Modern meat: Inside the slaughterhouse." Public Broadcasting Service. Accessed February 2015. http://www.pbs.org/wgbh/pages/frontline/shows/meat/slaughter/slaughterhouse.html.

Myers, B. R. "Slaughterhouse rules: A professor spends a season in hell." *The Atlantic*, October 24, 2012. Accessed February 2015. http://www.theatlantic.com/magazine/archive/2012/11/slaughterhouse-rules/309113/.

Painter, Stan, and Wenonah Hauter. Stan Painter and Wenonah Hauter to Ed Schafer. Washington, DC. February 12, 2008.

Pollan, Michael. *The Omnivore's Dilemma*. New York: Penguin Books, 2006.

"Questions about death of the animal when different stunning methods are used before Halal or Kosher slaughter." Dr. Temple Grandin's Web Page. Accessed February 2015. http://www.grandin.com/ritual/questions.diff.stun.methods.html.

"Rampant animal cruelty at California slaughter plant." The Humane Society of the United States. Last modified January 30, 2008. http://www.humanesociety.org/news/news/2008/01/undercover_investigation_013008.html.

Shames, Lisa. "Humane methods of handling and slaughter: Public reporting on violations can identify enforcement challenges and enhance transparency." US GAO. April 17, 2008. http://www.gao.gov/products/GAO-08-686T.

Shames, Lisa. "Humane Methods of Slaughter Act: Weaknesses in USDA enforcement." US GAO. March 4, 2010. http://www.gao.gov/assets/130/124119.pdf.

Sinclair, Upton. *The Jungle*. New York: Doubleday, 1906.

"Slaughter without stunning and food labeling." British Veterinary Association, 2013. http://www.bva.co.uk/uploadedFiles/Content/News,_campaigns_and_policies/Campaigns/Europe/Briefing_-_Welfare_at_slaughter_Oct_2013.pdf.

"Slaughterhouse workers." Food Empowerment Project. Accessed February 2015. http://www.foodispower.org/slaughterhouse-workers/.

Stull, Donald D., and Michael J. Broadway. *Slaughterhouse Blues: The Meat and Poultry Industry in North America.* Boston: Cengage Learning, 2003.

"Summary report for: 51-3023.00—slaughterers and meat packers," O*NET OnLine. Last modified 2010. http://www.onetonline.org/link/summary/51-3023.00.

Tolstoy, Leo. *The First Step: An Essay on the Morals of Diet.* Albert Broadbent, 1900.

United States Department of Agriculture Food Safety and Inspection Service. Accessed February 2015. http://www.fsis.usda.gov/wps/portal/fsis/home.

Welty, Jeff. "Humane slaughter laws." *Law and Contemporary Problems* 70 (2007): 175-206.

Withnall, Adam. "Denmark bans kosher and halal slaughter as minister says 'animal rights come before religion.'" *The Independent,* February 18, 2014. Accessed February 2015. http://www.independent.co.uk/news/world/europe/denmark-bans-halal-and-kosher-slaughter-as-minister-says-animal-rights-come-before-religion-9135580.html.

"Workplace safety and health: Safety in the meat and poultry industry, while improving, could be further strengthened." US GAO. January 2005. http://www.gao.gov/new.items/d0596.pdf.

Wyatt, Dean. "Continuing problems in USDA's enforcement of the Humane Methods of Slaughter Act." Domestic Policy Subcommittee Oversight and Government Reform Committee. March 4, 2010. http://www.gpo.gov/fdsys/pkg/CHRG-111hhrg65127/html/CHRG-111hhrg65127.htm.

"Your responsibilities under the Meat Regulation." Ontario Ministry of Agriculture, Food and Rural Affairs. Last modified November 13, 2013. http://www.omafra.gov.on.ca/english/food/inspection/meatinsp/resp-under-meat.htm.

Chapter Nine: The Wondrous Islands of Indonesia

"19th and 20th new human cases of avian influenza H5N1 in Cambodia in 2013." World Health Organization. Accessed February 2015. http://www.wpro.who.int/mediacentre/releases/2013/20130920/en/index.html.

Ackerman, Frank, and Elizabeth A. Stanton. "The Cost of Climate Change: What We'll Pay if Global Warming Continues Unchecked." National Resources Defense Council. May 2008. Accessed February 2015. http://www.nrdc.org/globalwarming/cost/cost.pdf. NRDC Global warming report_2008.

"Agriculture." Ontario Farm Animal Council. Accessed February 2015. http://www.farmsfoodfun.com/File.aspx?id=a3846369-ac79-42a3-a350-ce543a347241.

Ambarawati, Gusti Agung Ayu, et al. "The Cost to the Bali Beef Industry of the October 2002 Terrorist Attack." *Agribusiness Review* 12 (2004). Accessed February 2015. http://www.agrifood.info/review/2004/Ambarawati.html.

"Avian Influenza A/(H5N1) Cumulative Number of Confirmed Human Deaths." The Henry J. Kaiser Family Foundation. Accessed February 2015. http://kff.org/global-indicator/avian-flu-deaths/.

"Avian influenza A(H7N9) virus." World Health Organization. Accessed February 2015. http://www.who.int/influenza/human_animal_interface/influenza_h7n9/en/.

"Bali Cattle." Bali Cattle.com. Accessed February 2015. http://balicattle.com/.

"Climate Change in Indonesia." Global Greenhouse Warming. Accessed February 2015. http://www.global-greenhouse-warming.com/climate-change-in-Indonesia.html.

"Cock Fighting." Right Tourism. Accessed February 2015. http://right-tourism.com/issues/cruel-sports/cock-fighting/.

"Cumulative number of confirmed cases for avian influenza A(H5N1) reported to WHO, 2003-2013." World Health Organization. Accessed February 2015. http://www.who.int/influenza/human_animal_interface/EN_GIP_20130829CumulativeNumberH5N1cases.pdf.

"Cumulative number of confirmed human cases for avian influenza A(H1N1) reported to WHO, 2003-2014." World Health Organization. Accessed February 2015. http://www.who.int/influenza/human_animal_interface/EN_GIP_20140124CumulativeNumberH5N1cases.pdf.

Curry, Andrew. "Archaeology: The milk revolution." *Nature International Weekly Journal of Science.* July 31, 2013. Accessed February 2015. http://www.nature.com/news/archaeology-the-milk-revolution-1.13471.

Dawood, Fatimah S., et al. "Estimated global mortality associated with the first 12 months of 2009 pandemic influenza A H1N1 virus circulation: a modeling study." *The Lancet, Infectious Diseases* 12:9 (2012). Accessed February 2015. doi: http://dx.doi.org/10.1016/S1473-3099(12)70121-4.

"Dogs in Islam." IslamicConcern.com. Accessed February 2015. http://www.islamicconcern.com/dogs.asp.

"Fast Facts About Agriculture." The Voice of Agriculture. Accessed February 2015. http://www.fb.org/index.php?fuseaction=newsroom.fastfacts.

Fiala, Nathan. "How Meat Contributes to Global Warming." Scientific American. February 2009. Accessed February 2015. http://www.scientificamerican.com/article/the-greenhouse-hamburger/.

Field, Christopher, et al. "Climate Change 2014: Impacts, Adaptation, and Vulnerability." WGII AR5 Phase I Report Launch. March 31, 2014.

Foltz, Richard. *Animals in Islamic Traditions and Muslim Cultures.* Oxford, England: Oneworld Publication, 2006.

Geertz, Clifford. "Deep Play: Notes on the Balinese Cockfight." Accessed February 2015. http://www.rochester.edu/college/psc/clarke/214/Geertz72.pdf.

Gilbert, Elizabeth. *Eat Pray Love.* New York: Penguin, 2007.

"Global Alert and Response (GAR): Pandemic (H1N1) 2009." World Health Organization. Accessed February 2015. http://www.who.int/csr/disease/swineflu/en/.

Gordon, Michael, and Coral Davenport. "Kerry Implores Indonesia on Climate Change Peril." *New York Times*, February 16, 2014. Accessed February 2015. http://www.nytimes.com/2014/02/17/world/asia/kerry-urges-indonesia-to-help-stem-climate-change.html.

"H5N1 Avian Flu (H5N1 Bird Flu)." FLU.gov. Accessed February 2015. http://www.flu.gov/about_the_flu/h5n1/.

Hamilton, Jon. "Flu Fears Threaten Thai Cockfighting Tradition." *National Public Radio*, February 23, 2005. Accessed February 2015. http://www.npr.org/templates/story/story.php?storyId=4508912.

Hindey, Robin. "UCSF Epidemiologist Describes Similarities between Swine Flu and 1918 Killer." Accessed February 2015. http://www.ucsf.edu/news/2009/09/3064/swine-flu-shows-alarming-resemblance-1918-pandemic-ucsf-doc-says.

"INDONESIA: Avian flu deadliest in the world." IRIN humanitarian news and analysis. Accessed February 2015. http://www.irinnews.org/report/91387/indonesia-avian-flu-deadliest-in-the-world.

"Indonesia Country Profile." Oxfam New Zealand. Accessed February 2015. http://www.oxfam.org.nz/what-we-do/where-we-work/indonesia/indonesia-profile.

Naik, Zakir. "Questions Commonly Asked by Non-Muslims—IV: Why is Pork Forbidden?" *Islamic Voice; Religion* 13-02(146)(1999). Accessed February 2015. http://islamicvoice.com/february.99/zakir.htm.

Norrie, Justin. "Bird flu outbreaks blamed on rats." *The Sydney Morning Herald*, February 19, 2007. Accessed February 2015. http://www.smh.com.au/news/world/bird-flu-outbreaks-blamed-on-rats/2007/02/18/1171733612885.html.

"Number of confirmed human cases of avian influenza A(H7N9) reported to WHO: Report 9—data in WHO/HQ as of 12 August 2013, 14:45 GMT+1." World Health Organization. Accessed February 2015. http://www.who.int/influenza/human_animal_interface/influenza_h7n9/09_ReportWebH7N9Number.pdf.

Pachauri, R. K., et al. "Climate change 2014 Synthesis Report." IPCC Fifth Assessment Synthesis Report.

Potter, C.W. "A history of influenza." *Journal of Applied Microbiology* 91 (2001).

Riztyan, et al. "Genetic diversity and population structure of Indonesian native chickens based on single nucleotide polymorphism markers." *Oxford Journals; Poultry Science* 90(11)(2011): 2471-2478. Accessed February 2015. doi: 10.3382/ps.2011-01450.

Shurtleff, William, and Akiko Aoyagi. *History of Tempeh and Tempeh Products.* Lafayette: Soyinfo Center, 2011. Accessed February 2015. http://www.soyinfocenter.com/HSS/tempeh1.php.

Soedjana, Tjeppy D. "Impact and Implications of the Economic Crisis on Livestock Production, Marketing and Consumption in Indonesia." (paper presented at the workshop on the implications of the Asian economic crisis for the livestock industry, Bangkok, Thailand, August 1999). Accessed February 2015. http://www.fao.org/docrep/004/ac473e/AC473E04.htm.

Steckelberg, James M. "What's the difference between a bacterial infection and a viral infection?" Mayo Clinic. Accessed February 2015. http://www.mayoclinic.org/diseases-conditions/infectious-diseases/expert-answers/infectious-disease/faq-20058098.

Taubenberger, Jeffery K., and David M. Morens. "1918 Influenza: the Mother of All Pandemics." *Emerging Infectious Diseases* 12(1): 15-22. Accessed February 2015. doi: 10.3201/eid1201.050979.

Ver Berkmoes, Ryan, et al. *Lonely Planet Indonesia.* Victoria: Lonely Planet, 2009.
"Water Harvesting and Aquaculture for Rural Development: Introduction to Fish Culture in Rice Paddies." International Center for Aquaculture and Aquatic Environments Auburn University. Accessed February 2015. http://www.ag.auburn.edu/fish/documents/International_Pubs/Water%20Harvesting/English/Introduction%20to%20fish%20culture%20in%20rice%20paddies.pdf.

"WHO Risk Assessment: Human infections with avian influenza A(H7N9) virus." World Health Organization. Accessed February 2015. http://www.who.int/influenza/human_animal_interface/influenza_h7n9/riskassessment_h7n9_27june14.pdf?ua=1.

Chapter Ten: Fast Food Nation Malaysia

"About KFC." KFC. Accessed February 2015. http://www.kfc.com/about/.

"ACNeilsen Finds U.S. Truly is a Fast Food Nation; In International Study, U.S. Among the Leaders in Fast Food Dining, but Asia Tops List for Dining Frequency." Business Wire. Accessed February 2015. http://www.businesswire.com/news/home/20041221005644/en/ACNielsen-Finds-U.S.-Fast-Food-Nation-International#.VAjKofldWSo.

"Annual Report 2007." KFC Holdings (Malaysia) BHD. Accessed February 2015. http://cdn1.i3investor.com/my/files/dfg s88n/2008/04/08/1448850305-221479657.pdf.

"Annual Report 2013." Huat Lai Resources Berhad. Accessed February 2015. http://announcements.bursamalaysia.com/EDMS/subweb.nsf/7f04516f80986803 48256c6f0017a6bf/ca4e323854025c4c48257cec000ff2e6/$FILE/HUATLAI-AnnualReport2013.pdf.

"Big Dutchman: Poultry Production." Big Dutchman. Accessed February 2015. https://www.bigdutchman.com/en/poultry-production/news.html.

"Burger King aims to open more stores in 2012." *The Jakarta Post*, January 5, 2012. Accessed February 2015. http://www.thejakartapost.com/news/2012/01/05/burger-king-aims-open-more-stores-2012.html.

Butcher, Gary D., and Richard D. Miles. "Infectious Bursal Disease (Gumboro) in Commercial Broilers." *University of Florida IFAS Extension EDIS* 84 (2009). Accessed February 2015. http://edis.ifas.ufl.edu/vm024.

"Cargill." Cargill, Incorporated. Accessed February 2015. http://www.cargill.com/.

"Changing Structure of Global Food Consumption and Trade." U.S. Department of Agriculture. Accessed February 2015. http://www.ers.usda.gov/media/293645/wrs011_1_.pdf.

"Chicken & Turkey Meat Production in Canada Industry Market Research Report Now Available from IBISWorld." Digital Journal. Accessed February 2015. http://www.digitaljournal.com/pr/1517064.

"Chickens 'one-up' humans in ability to see color." Washington University School of Medicine. ScienceDaily. Accessed February 2015. http://www.sciencedaily.com/releases/2010/02/100216101159.htm.

"Cobb-Vantress, Inc." Cobb-Vantress, Inc. Accessed February 2015. http://www.cobb-vantress.com/.

"Cobb-Vantress, Inc." Vault. Accessed February 2015. http://www.vault.com/company-profiles/agriculture-forestry/cobb-vantress,-inc/company-overview.aspx.

"Complete Poultry Systems." Cumberland. Accessed February 2015. http://cumberlandpoultry.com/english/overview.html.

"Countries without McDonald's." *Fox News*, August 8, 2013. Accessed February 2015. http://www.foxnews.com/leisure/2013/08/08/countries-without-mcdonald/.

"Country Report: Fast Food in Malaysia." Euromonitor International. Accessed February 2015. http://www.euromonitor.com/fast-food-in-malaysia/report.

"Diseases and Conditions: E. coli." Mayo Clinic. Accessed February 2015. http://www.mayoclinic.org/diseases-conditions/e-coli/basics/definition/con-20032105.

"E.coli." Centers for Disease Control and Prevention. Accessed February 2015. http://www.cdc.gov/ecoli/.

"Facts About Palm Oil and Rainforests." Rainforest Rescue. Accessed February 2015. http://www.rainforest-rescue.org/topics/palm-oil.

"Fast-Food Supplier Suspends China Plant's Operations." *New York Times*, July 28, 2014. Accessed February 2015. http://www.nytimes.com/2014/07/29/business/fast-food-supplier-suspends-china-plants-operations.html?_r=0.

"Food and Beverage." Berjaya Corporation Berhad. Accessed February 2015. http://www.berjaya.com/food-beverage.html.

"Food Outlook; Global Market Analysis November 2012." Accessed February 2015. FAO Trade and Markets Division. http://www.fao.org/docrep/016/al993e/al993e00.pdf.

"Getting to Know Us." McDonald's. Accessed February 2015. http://www.aboutmcdonalds.com/mcd/our_company.html.

"Global Meat Production and Consumption Continue to Rise." Worldwatch Institute. Accessed February 2015. http://www.worldwatch.org/global-meat-production-and-consumption-continue-rise-1.

Hin, Cha Tien. "KFC Malaysia Sees Fastest Sales Growth in 10 Years." *Bloomberg*, December 3, 2008. Accessed February 2015. http://www.bloomberg.com/apps/news?pid=newsarchive&sid=aQyOxMPD9raw.

"How the Cobb 500 Changed the US Market." The Poultry Site. Accessed February 2015. http://www.thepoultrysite.com/articles/1200/how-the-cobb-500-changed-the-us-market.

"Huat Lai Resources Berhad." Huat Lai. Accessed February 2015. http://huatlai.com/.

"Infectious Bursal Disease (IBD): Causative Agent, Diagnosis and Prevention." Canadian Poultry Consultants Ltd. Accessed February 2015. http://www.canadianpoultry.ca/infectious_bursal_disease.htm.

"Infectious Bursal Disease, IBD, Gumboro." The Poultry Site. Accessed February 2015. http://www.thepoultrysite.com/diseaseinfo/81/infectious-bursal-disease-ibd-gumboro.

"Information on Avian Influenza." Centers for Disease Control and Prevention. Accessed February 2015. http://www.cdc.gov/flu/avianflu/.

"International Locations; North America, Latin America & Caribbean." Burger King Corporation. Accessed February 2015. http://www.bk.com/en/us/international/index.html.

"Key Facts About Avian Influenza (Bird Flu) and Highly Pathogenic Avian Influenza A (H5N1) Virus." Centers for Disease Control and Prevention. Accessed February 2015. http://www.cdc.gov/flu/avian/gen-info/facts.htm.

"KFC Malaysia." QSR Brands (M) Holdings Sdn Bhd. Accessed February 2015. http://www.kfc.com.my/kfc-malaysia/.

Kristof, Nicholas. "The Unhealthy Meat Market." *New York Times*, March 12, 2014. Accessed February 2015. http://www.nytimes.com/2014/03/13/opinion/kristof-the-unhealthy-meat-market.html.

Laman Utama. Accessed February 2015. http://www.mardi.gov.my/.

"Lohmann Tierzucht." Lohmann Tierzucht. Accessed February 2015. http://www.ltz.de/index.php.

Lubin, Gus, and Mamta Badkar. "15 Facts About McDonald's That Will Blow Your Mind." *Business Insider*, December 17, 2010. Accessed February 2015. http://www.businessinsider.com/amazing-facts-mcdonalds-2010-12?op=1.

"Malaysia Population and Country Demographics." Wonderful Malaysia. Accessed February 2015. http://www.wonderfulmalaysia.com/malaysia-population-and-demographics.htm.

"Malaysia vs. Indonesia" Find the Data. Accessed February 2015. http://country-facts.findthebest.com/compare/81-170/Malaysia-vs-Indonesia.

"McDonald's 'bigger than Jesus Christ'." *Marketing Week*, July 21, 1995. Accessed February 2015. http://www.marketingweek.co.uk/mcdonalds-bigger-than-jesus-christ/2012096.article.

"McDonald's Malaysia appoints new head." The Sundaily. Accessed February 2015. http://www.thesundaily.my/news/822037.

"Meat Consumption in China Now Double That in the United States." Janet Larson and Earth Policy Institute. Accessed February 2015. http://www.earth-policy.org/plan_b_updates/2012/update102.

"Newcastle Disease in Poultry (Avian pneumencephalitis, Exotic or velogenic Newcastle disease)." The Merck Veterinary Manual. Accessed February 2015. http://www.merckmanuals.com/vet/poultry/newcastle_disease_and_other_paramyxovirus_infections/newcastle_disease_in_poultry.html.

"Newcastle Disease Virus (NDV)." Avianbiotech.com. Accessed February 2015. http://www.avianbiotech.com/Diseases/Newcastle.htm.

"Obesity and Overweight." Centers for Disease Control and Prevention. Accessed February 2015. http://www.cdc.gov/nchs/fastats/obesity-overweight.htm.

"OECD-FAO Agricultural Outlook 2013-2022." Organization for Economic Co-operation and Development and the Food and Agriculture Organization. Accessed February 2015. http://www.keepeek.com/Digital-Asset-Management/oecd/agriculture-and-food/oecd-fao-agricultural-outlook-2013_agr_outlook-2013-en#page184.

"OECD-FAO Agricultural Outlook 2014-2023." Organization for Economic Co-operation and Development and the Food and Agriculture Organization. Accessed February 2015. http://www.oecd.org/site/oecd-faoagriculturaloutlook.

"Oil palm plantations: threats and opportunities for tropical ecosystems." United Nations Environment Programme. Accessed February 2015. http://www.unep.org/pdf/Dec_11_Palm_Plantations.pdf.

"Overview of Infectious Bursal Disease in Poultry." The Merck Veterinary Manual. Accessed February 2015. http://www.merckmanuals.com/vet/poultry/infectious_bursal_disease/overview_of_infectious_bursal_disease_in_poultry.html.

"Poultry Breeds." Comercial Solana Global S.L. Accessed February 2015. http://www.chickens-and-chicken-eggs.com/poultry-breeds.html.

"Poultry Integration." KFC Holdings (Malaysia) Bhd. Accessed February 2015. http://www.pocketzila.com/kfch/business-units/poultry-integration/.

Raghavan, V. "Concentration and Integration." *World Poultry—Elsevier* 18(8) (2002). Accessed February 2015. http://www.worldpoultry.net/PageFiles/25997/001_boerderij-download-WP6296D01.pdf.

Richmond, Simon, et al. *Lonely Planet Malaysia, Singapore & Brunei*. Victoria: Lonely Planet, 2013.

"Sanovo Technology Group." Sonovo Technology Netherlands. Accessed February 2015. http://www.staalkat.com/.

Schlosser, Eric. *Fast Food Nation*. New York: Houghton Mifflin, 2001.

"The Science of Poultry Lighting. A Bird's Eye View." Once Innovations. Accessed February 2015. http://www.onceinnovations.com/wp/wp-content/uploads/2014/10/Science-of-Poultry-Vision-Single-Pages1.pdf.

Shields, Sara, and Michael Greger. "Animal Welfare and Food Safety Aspects of Confining Broiler Chickens to Cages." *Animals* 3(2) (2013): 386-400. Accessed February 2015. doi: 10.3390/ani3020386.

"Statistics about E-coli food poisoning." Right Diagnosis from healthgrades. Accessed February 2015. http://www.rightdiagnosis.com/e/e_coli_food_poisoning/stats.htm.

Steinfeld, Henning, et al. *Livestock's long shadow: environmental issues and options.* Rome: Food and Agriculture Organization of the United Nations, 2006. http://www.europarl.europa.eu/climatechange/doc/FAO%20report%20executive%20summary.pdf.

"Tecno, the eggsperience." Tecno Poultry Equipment Spa. Accessed February 2015. http://www.poultryequipment.com/en/.

"Top 10's Crème de la Crème Food Brands In Malaysia." Top 10 of Malaysia. Accessed February 2015. http://top10malaysia.com/home/index.php/news-and-events/top-10-food-brands-in-malaysia.

Triall, Bruce. "The double burden of malnutrition; Case studies from six developing countries." *Food and Nutrition Paper* 84 (2006). http://www.fao.org/docrep/009/a0442e/a0442e0x.htm.

"UNEP Global Environmental Alert Service (GEAS)." United Nations Environment Programme. Accessed February 2015. http://www.unep.org/pdf/Dec_11_Palm_Plantations.pdf.

"The Wendy's Company." Wendy's International. Accessed February 2015. http://www.aboutwendys.com/Our-Company/.

"What is Palm Oil?" Green Palm Sustainability. Accessed February 2015. http://www.greenpalm.org/en/about-palm-oil/what-is-palm-oil-used-in.

"World Agriculture Towards 2030/2050." ESA Working Paper No. 12-03. Food and Agriculture Organization of the United States. June 2012. Accessed February 2015. http://www.fao.org/docrep/016/ap106e/ap106e.pdf.

"Yum!'s Global Restaurant System." Yum Brands RSC. Accessed February 2015. http://www.yum.com/company/map.asp.

Chapter Eleven: Lion City Singapore

"A Brief History of Singapore." Your Singapore. Accessed February 2015. http://www.yoursingapore.com/content/traveller/en/browse/aboutsingapore/a-brief-history.html.

Adams, Williams, and Damian Ma. "Appetite for Destruction." *Foreign Policy*, October 1, 2013. Accessed February 2015. http://foreignpolicy.com/2013/10/01/appetite-for-destruction/.

Brown, Felicity. "Meat Consumption Per Capita." *The Guardian*, September 2, 2009. Accessed February 2015. http://www.theguardian.com/environment/datablog/2009/sep/02/meat-consumption-per-capita-climate-change.

"Country Facts Side-by-Side: Malaysia vs. Indonesia." Find the Data. Accessed February 2015. http://country-facts.findthebest.com/compare/81-170/Malaysia-vs-Indonesia.

"Current Worldwide Annual Meat Consumption Per Capita." ChartsBin. Accessed February 2015. http://chartsbin.com/view/12730.

"Global Meat Production and Consumption Continue to Rise." Worldwatch Institute. Last modified October 11, 2011. Accessed February 2015. http://www.worldwatch.org/global-meat-production-and-consumption-continue-rise-1.

"International Egg and Poultry Review." The Poultry Site. October 5, 2004. Accessed February 2015. http://www.thepoultrysite.com/poultrynews/7037/international-egg-and-poultry-review.

"Meat Consumption in China Now Double That in the United States." Earth Policy Institute. Last modified April 24, 2012. http://www.earth-policy.org/plan_b_updates/2012/update102.

"Meat Consumption per Person." Scribd. Last modified May 2, 2012. http://www.scribd.com/EconomistDailychart/d/91840616-Meat-Consumption-Per-Person.

"OECD-FAO Agricultural Outlook 2013-2022." OECD-FAO. June 27, 2013. Accessed February 2015. doi: 10.1787/agr_outlook-2013-en

"Poll: Are Singaporeans Least Happy?" *Cable News Network,* December 21, 2012. Accessed February 2015. http://www.cnn.com/2012/12/21/world/asia/singapore-least-happy/.

Seng Choon Farm. Accessed February 2015. http://www.sengchoonfarm.com/.

Steinfeld, Henning, et al. *Livestock's long shadow: environmental issues and options.* Rome: Food and Agriculture Organization of the United Nations, 2006. http://www.europarl.europa.eu/climatechange/doc/FAO%20report%20executive%20summary.pdf.

"World Agriculture Towards 2030/2050." ESA Working Paper No. 12-03. Food and Agriculture Organization of the United States. June 2012. Accessed February 2015. http://www.fao.org/docrep/016/ap106e/ap106e.pdf.

"The World Factbook Country Comparison: Total Fertility Rate." Central Intelligence Agency. Accessed February 2015. https://www.cia.gov/library/publications/the-world-factbook/rankorder/2127rank.html.

Chapter Twelve: The Monsters of Mexico

"Are We Becoming China's Factory Farm?" Mother Jones. December 2013. Accessed February 2015. http://www.motherjones.com/media/2014/03/china-factory-farm-america-pork.

Bunge, Jacob, and Michael Calia. "Tyson Agrees to Sell Latin America Chicken Assets." *Wall Street Journal,* July 28, 2014. Accessed February 2015. http://online.wsj.com/articles/tysons-profit-rises-but-misses-views-1406543524.

"Chinese Agribusiness Profiles: Shineway Group." Pig Penning Blog. Accessed February 2015. http://pigpenning.wordpress.com/agbiz-profiles/shineway-shuanghui-group/.

Chore-Time Poultry Production System. Accessed February 2015. http://choretimepoultry.com/.

Costco. Accessed February 2015. http://www3.costco.com.mx/wps/portal/publico/SUCURSALES.

Crio. Accessed February 2015. http://www.crio.com.mx/.

Dawood, Fatimah S., et al. "Estimated global mortality associated with the first 12 months of 2009 pandemic influenza A H1N1 virus circulation: a modeling study." *The Lancet, Infectious Diseases* 12:9 (2012). Accessed February 2015. doi: http://dx.doi.org/10.1016/S1473-3099(12)70121-4.

Domino's: Mexico. Accessed February 2015. http://dominos.com.mx/quienes-somos/.

"Environment." Grace Communications Foundation. Accessed February 2015. http://www.sustainabletable.org/265/environment.

"EU Egg Production Beyond the 2012 Cage Ban." The Poultry Site. Last modified December 9, 2009. http://www.thepoultrysite.com/articles/1576/eu-egg-production-beyond-the-2012-cage-ban.

"Examining Mexico's Protein Consumption, Import Trends." U.S. Meat Export Federation. Accessed February 2015. https://www.usmef.org/examining-mexicos-protein-consumption-import-trends/.

"Facts About Pollution From Livestock Farms." National Resources Defense Council. Last modified February 21, 2013. Accessed February 2015. http://www.nrdc.org/water/pollution/ffarms.asp.

Fainaru, Steve. "Mexicans Blame Industrial Hog Farms." *Washington Post*, May 10, 2009. Accessed February 2015. http://www.washingtonpost.com/wp-dyn/content/article/2009/05/09/AR2009050902531.html.

Fuhrman, Monte W. "Hog Heaven: Is it on the Yucatan Peninsula of Mexico?" *Journal of Swine Health and Production* vol. 4, no. 1 (1996): 41-43.

Granjas Carroll de Mexico. Accessed February 2015. http://www.granjascarroll.mx/ing_gcm_somos.php.

"Hog Wild: Factory Farms are Poisoning Iowa's Drinking Water." OnEarth. Last modified February 24, 2014. Accessed February 2015. http://www.onearth.org/articles/2014/02/factory-farms-are-poisoning-iowa-water.

"How to Manage Manure." Healthy Landscapes. Accessed February 2015. http://www.uri.edu/ce/healthylandscapes/livestock/how_manure_overall.htm.

"Huge Spill of Hog Waste Fuels an Old Debate in North Carolina." *New York Times*, June 25, 1995. Accessed February 2015. http://www.nytimes.com/1995/06/25/us/huge-spill-of-hog-waste-fuels-an-old-debate-in-north-carolina.html.

Keken. Accessed February 2015. http://www.keken.com.mx/index.php.

Kristof, Nicholas. "Abusing Chickens We Eat." *New York Times*, December 3, 2014. Accessed February 2015. http://www.nytimes.com/2014/12/04/opinion/nicholas-kristof-abusing-chickens-we-eat.html?_r=0.

"Kuo 2013 Annual Report." Kuo Companies. Accessed February 2015. http://www.kuo.com.mx/anexos/eng/kuo_2013_ingles_1pag.pdf.

"Mad Cow California: Stop Feeding Cows Chicken Manure." *Huffington Post*, April 26, 2012. Accessed February 2015. http://www.huffingtonpost.com/michael-greger-md/mad-cow-disease-california_b_1450994.html.

McDonald's: Mexico. Accessed February 2015. www.mcdonalds.com.mx/.

"Mexico Starts Pork Exports to China." The Pig Site. Last modified November 2, 2012. Accessed February 2015. http://www.thepigsite.com/swinenews/31374/mexico-starts-pork-exports-to-china.

Murphy, Ryan. "Can Smithfield Meet Chinese Demand Without Driving Up U.S. Prices?" *Daily Press*, September 29, 2013. Accessed February 2015. http://articles.dailypress.com/2013-09-29/news/dp-nws-smithfield-sale-follow-20130929_1_daniel-slane-smithfield-foods-shuanghui-international-holdings.

"National Pork Board Examines International Marketing Opportunities." Pork Checkoff. Last modified August 14, 2014. Accessed February 2015. http://www.pork.org/national-pork-board-examines-international-marketing-opportunities/.

"Overview of Sudden Death Syndrome of Broiler Chickens." The Merck Veterinary Manual. Last modified August 2013. Accessed February 2015. http://www.

merckmanuals.com/vet/poultry/sudden_death_syndrome_of_broiler_chickens/overview_of_sudden_death_syndrome_of_broiler_chickens.html.

Pei, Minxin. "The Real Reason Behind Shuanghui's Purchase of Smithfield." Fortune. June 3, 2013. Accessed February 2015. http://fortune.com/2013/06/04/the-real-reason-behind-shuanghuis-purchase-of-smithfield/.

Pollan, Michael. *The Omnivore's Dilemma*. New York: Penguin Books, 2006.

"Pork Production: Potential Environmental Impacts of Animal Feeding Operations." Environmental Protection Agency. Accessed February 2015. http://www.epa.gov/agriculture/ag101/printpork.html#impact.

"Q&A With Temple Grandin Part II." Food Safety News. June 23, 2010. Accessed February 2015. http://www.foodsafetynews.com/2010/06/qa-with-temple-grandin-part-ii/#.Ul7VVFCkeSo.

"RG-2 Regulatory Guidance: Feeding of Poultry Manure to Cattle Prohibited." Canadian Food Inspection Agency. Last modified October 9, 2014. http://www.inspection.gc.ca/animals/feeds/regulatory-guidance/rg-2/eng/1328859106165/1328859287377.

"Shuanghui International and Smithfield Foods Complete Strategic Combination, Creating a Leading Global Pork Enterprise." Smithfield. September 26, 2013. Accessed February 2015. http://investors.smithfieldfoods.com/releasedetail.cfm?releaseid=793522.

Stan County Planning Board: Swine Waste Removal. Accessed February 2015. http://www.stancounty.com/planning/pl/agenda/2009/10-01-09/RWQCB%20Part%204.pdf.

"Swine Flu Outbreak Could be Linked to Smithfield Factory Farms." Grist. April 26, 2009. Accessed February 2015. http://grist.org/article/2009-04-25-swine-flu-smithfield/.

"U.S. Broiler Performance: 1925 to Present." National Chicken Council. Accessed February 2015. http://www.nationalchickencouncil.org/about-the-industry/statistics/u-s-broiler-performance/.

"U.S. Cattle Being Fed Chicken Manure?" The Bovine Blog. Last modified June 1, 2009. Accessed February 2015. http://thebovine.wordpress.com/2009/06/01/u-s-cattle-being-fed-chicken-manure/.

"U.S. Exports Continue to See Strong Growth in Asia." Global Meat News. Last modified August 11, 2014. Accessed February 2015. http://www.globalmeatnews.com/Regions/Asia/South-Korea/US-exports-continue-to-see-strong-growth-in-Asia.

Walmart: Mexico. Accessed February 2015. http://www.walmart.com.mx/.

"Why Invest in Maple Leaf Foods?" Maple Leaf. Accessed February 2015. http://investor.mapleleaf.ca/phoenix.zhtml?c=88490&p=irol-homeprofile.

Wideman, R.F., D.D. Rhoads, G.F. Erf, and N.B. Anthony. "Pulmonary Arterial Hypertension (Ascites Syndrome) in Broilers: A Review." *Poultry Science* (2013), 92(1): 64. Accessed February 2015. http://ps.oxfordjournals.org/content/92/1/64.full.pdf+html.

"Yum! Financial Data." Yum! Brands. Accessed February 2015. http://www.yum.com/investors/restcounts.asp.

Chapter Thirteen: Barefoot in Beautiful Belize

"2007 Census of Agriculture: Women Farmers." United States Department of Agriculture. Accessed February 2015. http://www.agcensus.usda.gov/Publications/2007/Online_Highlights/Fact_Sheets/Demographics/women.pdf.

"2013 Food and Health Survey." International Food Information Council Foundation. Accessed February 2015. http://www.foodinsight.org/articles/2013-food-and-health-survey.

"About Us." Maple Leaf Foods. Accessed February 2015. http://www.mapleleaffoods.com/about-us/our-people-and-values.

Adams, Tim. "Testosterone and high finance do not mix: so bring on the women." *New York Times*, June 18, 2011. Accessed February 2015. http://www.theguardian.com/world/2011/jun/19/neuroeconomics-women-city-financial-crash.

"Agriculture—Pillar of the Belizean Economy." Ministry of Agriculture & Fisheries. Accessed February 2015. http://www.agriculture.gov.bz/Agriculture_Data.html.

Alfano, Sean. "How And Where America Eats." *CBS News*, November 20, 2005. Accessed February 2015. http://www.cbsnews.com/2100-500160_162-1060315.html.

"All About Birds: Cattle Egret." The Cornell Lab of Ornithology. Accessed February 2015. http://www.allaboutbirds.org/guide/cattle_egret/lifehistory.

"An HSUS Report: Food Safety and Cage Egg Production." The Humane Society of the United States. Accessed February 2015. http://www.humanesociety.org/assets/pdfs/farm/report_food_safety_eggs.pdf.

"Annual Report 2008; Agriculture: Pillar of the Belizean Economy." Ministry of Agriculture & Fisheries. Accessed February 2015. http://www.agriculture.gov.bz/PDF/Annual%20Report%202008.pdf.

"Attitudes of consumers towards the welfare of farmed animals Wave 2." *Special Eurobarometer* 229(2) (2007): Wave 64.4. Accessed February 2015. http://ec.europa.eu/food/animal/welfare/survey/sp_barometer_fa_en.pdf.

"Attitudes of EU citizens toward Animal Welfare." *Special Eurobarometer* 270 (2007): Wave 66.1. Accessed February 2015. http://ec.europa.eu/public_opinion/archives/ebs/ebs_270_en.pdf.

"Belize." The Nature Conservancy. Accessed February 2015. http://www.nature.org/ourinitiatives/regions/centralamerica/belize/.

"Belize Flag and National Symbols." Belize.com. Accessed February 2015. http://www.belize.com/belize-flag.

"Belizean Food." Belize.com. Accessed February 2015. http://www.belize.com/belize-fast-food.

Booth, Tom, and Paul Irwing. "The Distance Between Mars and Venus: Measuring Global Sex Differences in Personality." *PLOS One* 7(1) (2012): e29265. Accessed February 2015. doi: 10.1371/journal.pone.0029265. http://www.plosone.org/article/info%3Adoi%2F10.1371%2Fjournal.pone.0029265.

Chanona, Janelle. "Move Over Dala Chicken; Subway Opens in Belize." *News 5*, June 3, 2003. Accessed February 2015. http://edition.channel5belize.com/archives/15088.

Churchill, Winston. "Their Finest Hour" (speech in the House of Commons). June 18, 1940. Accessed February 2015. http://www.winstonchurchill.org/resources/speeches/1940-the-finest-hour/their-finest-hour.

Cohan, William D. "Does Wall Street Need an Estrogen Injection?" *New York Times*, April 1, 2010. Accessed February 2015. http://opinionator.blogs.nytimes.com/2010/04/01/does-wall-street-need-an-estrogen-injection/?_r=0.

Costa, Paul T. Jr., Antonio Terracciano, and Robert R. McCrae. "Gender Differences in Personality Traits Across Cultures: Robust and Surprising Findings." *Journal of*

Personality and Social Psychology 81(2) (2001): 322-331. Accessed February 2015. doi: 10.1037//0022-3514.81.2.322.

"Culture; Mennonites." Northernbelize.com. Accessed February 2015. http://www.northernbelize.com/cult_mennonite.html.

Doering, Christopher. "Breaking the 'grass ceiling': More women are farming." *USA Today*, March 17, 2013. Accessed February 2015. http://www.usatoday.com/story/news/nation/2013/03/17/women-farmers-increasing/1993009/.

"Facts About Belize." 2013 Belize Tourism Board. Accessed February 2015. https://www.travelbelize.org/facts-about-belize.

"Farmers Markets and Direct-to-Consumer Marketing." United States Department of Agriculture. Accessed February 2015. http://www.ams.usda.gov/AMSv1.0/farmersmarkets.

Harrison, Erin E. "The legal industry tackles gender diversity." Inside Counsel. September 8, 2014. Accessed February 2015. http://www.insidecounsel.com/2014/09/08/the-legal-industry-tackles-gender-diversity.

"Highlights of Women's Earnings in 2010." U.S. Department of Labor and U.S. Bureau of Labor Statistics. July 2011. Accessed February 2015. http://www.bls.gov/cps/cpswom2010.pdf.

"The History of Farmers Markets." Redmond Farmers Market. Accessed February 2015. http://www.redmondfarmersmarket.com/history-of-farmers-market.html.

Kristof, Nicholas. "Mistresses of the Universe." *New York Times*, February 7, 2009. Accessed February 2015. http://www.nytimes.com/2009/02/08/opinion/08kristof.html?_r=0.

"Leadership Team." Cal-Maine Foods, Inc. Accessed February 2015. http://www.calmainefoods.com/company/leadership-team.aspx.

"Leading the Way for Good Food." Smithfield. Accessed February 2015. http://www.smithfieldfoods.com/our-company/leadership/.

Lewis, Michael. "The Big Short: Inside the Doomsday Machine." *YouTube* video, posted by Berkeley Haas School. September 29, 2010. Accessed February 2015. http://www.youtube.com/watch?v=stnGC9jL8Fk.

Lewis, Michael. "Wall Street on the Tundra." *Vanity Fair.* April 2009. Accessed February 2015. http://www.vanityfair.com/culture/2009/04/iceland200904.

"Men or Women: Who's the Better Leader? A Paradox in Public Attitudes." Pew Research Center Social & Demographic Trends. Accessed February 2015. http://www.pewsocialtrends.org/2008/08/25/men-or-women-whos-the-better-leader/.

Micik, Katie. "Changing Face of Ag Women—1; Women Take More Active Production Role on Farm." *The Progressive Farmer,* October 27, 2011.

Moore, David W. "Public Lukewarm on Animal Rights." *Gallup Poll News Service,* May 21, 2003. Accessed February 2015. http://www.gallup.com/poll/8461/public-lukewarm-animal-rights.aspx.

Murphy, Clare. "It's official: men like meat." *BBC News,* March 19, 2008. Accessed February 2015. http://news.bbc.co.uk/2/hi/health/7305505.stm.

"Perspectives: Farmers Markets." Nourish. Accessed February 2015. http://www.nourishlife.org/2011/03/farmers_markets/.

Roberto E. Mercadillo, et al. "Perception of suffering and compassion experience: Brain gender disparities." *Brain and Cognition* (2011). Accessed February 2015. doi:10.1016/j.bandc.2011.03.019.

Rueckert, Linda, and Nicolette Naybar, "Gender differences in empathy: The role of the right hemisphere." *Brain and Cognition* 67 (2008): 162-167. Accessed February 2015. doi:10.1016/j.bandc.2008.01.002.

Smith, Betsy Brantner. "Women in law enforcement: How gender differences enable new police leaders." *PoliceOne.com*, May 17, 2011. Accessed February 2015. http://www.policeone.com/women-officers/articles/3680757-Women-in-law-enforcement-How-gender-differences-enable-new-police-leaders/.

"Statistics Explained." European Commission Eurostat. Accessed February 2015. http://epp.eurostat.ec.europa.eu/statistics_explained/index.php/Farm_structure_statistics.

"Tyson Foods Executive Team." Tyson Foods, Inc. Accessed February 2015. http://www.tysonfoods.com/Our-Story/Leadership/Executive-Bios.aspx.

"Unique Meeting Back in Time." The Family Without Borders. Accessed February 2015. http://thefamilywithoutborders.com/mennonites_in_belize-2012-05-14/.

Vecchio, Riccardo. "European and United States farms' markets: similarities, differences, and potential developments" (paper presented at the 113th EAAE Seminar "A resilient European food industry and food chain in a challenging world"). September 3-6, 2009. Accessed February 2015. http://ageconsearch.umn.edu/bitstream/58131/2/Vecchio.pdf.

Wilson, E.O. *On Human Nature*. Cambridge, MA: Harvard University Press, 2004.

"The World Factbook: Central America and Caribbean: Belize." Central Intelligence Agency. Accessed February 2015. https://www.cia.gov/library/publications/the-world-factbook/geos/bh.html.

Zogby, Joseph. "Nationwide Views On The Treatment of Farm Animals." *Zogby International*, October 22, 2003. Accessed February 2015. http://civileats.com/wp-content/uploads/2009/09/AWT-final-poll-report-10-22.pdf.

Chapter Fourteen: Ag-Gag America

"Ag 101: Dairy Production Systems." EPA. Accessed February 2015. http://www.epa.gov/oecaagct/ag101/dairysystems.html.

"Ag-Gag Bills at the State Level." American Society for the Prevention of Cruelty to Animals. Accessed February 2015. http://www.aspca.org/fight-cruelty/advocacy-center/ag-gag-whistleblower-suppression-legislation/ag-gag-bills-state-level.

"An HSUS Report: The Welfare of Cows in the Dairy Industry." The Humane Society of the United States.

"Anaerobic Digesters." Energy Justice Network. Accessed February 2015. http://www.energyjustice.net/digesters/.

Animal Legal Defense Fund. "ADLF Spearheads Legislation to Oppose Ag Gag Legislation." December 21, 2011. Accessed February 2015. http://www.aldf.org/article.php?id=1904.

"CA Economy." Netstate. Accessed February 2015. http://www.netstate.com/economy/ca_economy.htm.

"Canadian Dairy Breeds." CowPower. February 24, 2014. Accessed February 2015. http://www.cowpowerbc.com/press/canadian-dairy-breeds.

"California Dairy Statistics: 2011 Data." California Dairy Farmers Association. Accessed February 2015. http://www.cdfa.ca.gov/dairy/pdf/Annual/2011/2012_Annual_2011_Data.pdf.

Carson, Michael. "Earl Butz." *The Guardian*, February 4, 2008. Accessed February 2015. http://www.theguardian.com/world/2008/feb/04/usa.obituaries.

Carson, Rachel. *Silent Spring*. New York: Houghton Mifflin, 1962.

"Cattle, Sheep, and Goat Ringworm." Aardora. Accessed February 2015. http://www.aardora.com/animal-health/farm-skin-problems/ringworm.

"Changes in the Size and Location of U.S. Dairy Farms." U.S. Department of Agriculture. Accessed February 2015. http://www.ers.usda.gov/media/430528/err47b_1_.pdf.

"Corn, Cows, Feedlots, and Your Health." OnlyGrassFed. Accessed February 2015. http://www.onlygrassfed.com/pasture-vs-feedlot.html.

"Cruelty to Farm Animals Demands Exposure." *Washington Post*, April 26, 2013. Accessed February 2015. http://www.washingtonpost.com/opinions/cruelty-to-farm-animals-demands-exposure/2013/04/26/9a972c8e-a6bf-11e2-a8e2-5b98cb59187f_story.html.

"Dairy Facts." Purdue University Food Animal Education Network. Accessed February 2015. http://www.ansc.purdue.edu/faen/dairy%20facts.html.

Derbyshire, David. "Up to 16,000 hens crammed into a shed, and many never see daylight. But believe it or not, THESE are free-range chickens." *Daily Mail*, November 13, 2013. Accessed February 2015. http://www.dailymail.co.uk/news/article-2508173/16-000-free-range-chickens-crammed-shed-NEVER-daylight.html.

"Eating With Our Eyes Closed." *New York Times*, April 9, 2013. Accessed February 2015. http://www.nytimes.com/2013/04/10/opinion/eating-with-our-eyes-closed.html.

"Economics of Anaerobic Digesters for Processing Animal Manure." Extension. June 21, 2013. Accessed February 2015. http://www.extension.org/pages/19461/economics-of-anaerobic-digesters-for-processing-animal-manure.

"Facts About Farmworkers." National Center for Farmworker Health, Inc. August 2012. Accessed February 2015. http://www.ncfh.org/docs/fs-Facts%20about%20Farmworkers.pdf.

"Facts about Holstein Cattle." Holstein Association U.S.A. Accessed February 2015. http://www.holsteinusa.com/holstein_breed/holstein101.html?tab=2#TabbedPanels1.

Flagg, Kathryn. "Farmers look to alternative energy: science converts cow manure to profit for dairies." *Addison County Independent*, August 17, 2009. Accessed February 2015. http://www.addisonindependent.com/200908farmers-look-alternative-energy-science-converts-cow-manure-profit-dairies.

Flynn, Dan. "Five States Now Have 'Ag-Gag' Laws on the Books." *Food Safety News*, March 26, 2012. Accessed February 2015. http://worldnewsforlife.wordpress.com/2012/01/16/us-ag-gag-laws-pulling-the-curtain-on-the-cruelty-of-animal-farming/.

"General U.S. Stats." United Egg Producers. Last modified February 2015. Accessed February 2015. http://www.unitedegg.org/GeneralStats/default.cfm.

Gliona, John M. "Cow tail, cow pie: Fury in Colorado over possible 'docking' ban." *Los Angeles Times*, March 7, 2013. Accessed February 2015. http://articles.latimes.com/2013/mar/07/nation/la-na-nn-colorado-cow-tail-docking-20130307.

Hernandez-Mendo, O., M.A.G. von Keyserlingk, D.M. Veira, and D.M. Weary. "Effects of Pasture on Lameness in Dairy Cows." *Journal of Dairy Science* 90 (2007): 1209–1214.

Hubbs, Philip D. "The Origins and Consequences of the American Feedlot System." PhD diss., Baylor University, 2010.

"Industrial Livestock Production." Grace Communications Foundation. Accessed February 2015. http://www.sustainabletable.org/859/industrial-livestock-production.

"Industry Statistics: Milk Production (Most Recently) by State." StateMaster. Accessed February 2015. http://www.statemaster.com/graph/ind_mil_pro-industry-milk-production.

Jones, G.M., and T.L. Bailey, Jr. "Understanding the Basics of Mastitis." Virginia Cooperative Extension. Accessed February 2015. http://pubs.ext.vt.edu/404/404-233/404-233.html.

Lazarus, William F. "Farm-Based Anaerobic Digesters as an Energy and Odor Control Technology." U.S. Department of Agriculture. February 2008. Accessed February 2015. http://www.usda.gov/oce/reports/energy/AnerobicDigesters0308.pdf.

Lymbery, Phillip. *Farmaggedon: The True Cost of Cheap Meat*. London: Bloomsbury Publishing, 2014.

Maroney, Jr., James H. "Manure Digesters May Lower Farmers' Energy Costs, But They Do Little to Help the Environment." VT Digger. August 12, 2013. Accessed February 2015. http://vtdigger.org/2013/08/12/maroney-manure-digesters-may-lower-farmers-energy-costs-but-they-do-little-to-help-the-environment/.

Matsuoka, T., O.A. Muenster, E.E. Ose, and L. Tonkinson, "Orally administered tylosin for the control of pneumonia in neonatal calves." *Veterinary Record* 107 (1980): 149.

"More Vt dairy farms close." WCAX. April 11, 2012. Accessed February 2015. http://www.wcax.com/story/17383247/more-vt-dairy-farms-close.

Morrissey, Tim. "Dairy Cow Tail-Docking Called 'Inhumane Mutilation'." *Public News Service*, January 8, 2014. Accessed February 2015. http://www.publicnewsservice.org/2014-01-08/animal-welfare/dairy-cow-tail-docking-called-inhumane-mutilation/a36708-1.

Morter, R.L., and C. James Callahan. "Ringworm in Cattle." Animal Health. Accessed February 2015. https://www.extension.purdue.edu/extmedia/VY/VY-56.html.

"The National Agricultural Workers Survey." U.S. Department of Labor. Accessed February 2015. http://www.doleta.gov/agworker/report9/chapter1.cfm.

Pelzer, Kevin D., and Nancy Currin. "Zoonotic Diseases of Cattle." May 1, 2009. http://pubs.ext.vt.edu/400/400-460/400-460.html.

Podhaizer, Suzanne. "The Veal Deal." *Seven Days*, August 27, 2008. Accessed February 2015. http://www.7dvt.com/2008veal-deal.

Pollan, Michael. "Farmer in Chief." *New York Times Magazine*, October 9, 2008. Accessed February 2015. http://www.nytimes.com/2008/10/12/magazine/12policy-t.html?pagewanted=all&_r=0.

Pollan, Michael. *The Omnivore's Dilemma*. New York: Penguin Books, 2006.

"Protecting America From Foot-and-Mouth Disease and Other High-Consequence Livestock Diseases." APHIS: U.S. Department of Agriculture. July 2013. Accessed February 2015. http://www.aphis.usda.gov/publications/animal_health/2013/fs_fmd_hcd_protection.pdf.

Quigley, Jim. "Calf Note # 77: Methods of Feeding Water." CalfNotes. November 18, 2001. Accessed February 2015. http://www.calfnotes.com/pdffiles/CN077.pdf.

"Reducing Lameness in Dairy Cows." The Dairy Site. May 9, 2008. Accessed February 2015. http://www.thedairysite.com/articles/1430/reducing-lameness-in-dairy-cows.

"Ringworm." Cattle. Accessed February 2015. http://www.cattle.com/articles/title/Ringworm.aspx.

"Ringworm in Cattle." The Cattle Site. Accessed February 2015. http://www.thecattlesite.com/diseaseinfo/233/ringworm-in-cattle.

"Searching for Value in the Vermont Dairy Landscape." Sohl Photography. Accessed February 2015. http://www.sohlphotography.com/images/Vermont_Dairy_Sohl.pdf.

Sechen, Suzanne. "Bovine Somatotropin (bST)—Possible Increased Use of Antibiotics to Treat Mastitis in Cows." Food and Drug Administration, Center for Veterinary Medicine. October 30, 2013. Accessed February 2015. http://www.fda.gov/downloads/AnimalVeterinary/SafetyHealth/ProductSafetyInformation/UCM383073.pdf.

Shearer, J.K.. and S. Van Amstel. "Lameness in Dairy Cattle." Healthy Hooves. Accessed February 2015. http://healthyhooves.com/pdffiles/dr%20shearer.pdf.

Shriver, W. Gregory, et al. "The Distribution and Abundance of Obligate Grassland Birds Breeding in New England and New York." U.S. Department of Agriculture. 2005. Accessed February 2015. http://www.fs.fed.us/psw/publications/documents/psw_gtr191/psw_gtr191_0511-0518_shriver.pdf.

"Skin Problems in Young Cattle: Warts and Ringworm." *Tri-State Livestock News*, November 20, 2012. Accessed February 2015. http://www.tsln.com/home/3206858-111/warts-animal-cattle-ringworm.

Smith, S.E. "Agriculture Gag Laws are Violating Press Freedom in the U.S." *The Guardian*, June 6, 2012. Accessed February 2015. http://www.theguardian.com/commentisfree/2012/jun/06/agriculture-gag-laws-press-freedom.

"State Agricultural Profiles: Pacific Southwest Region 9." US Environmental Protection Agency. Accessed February 2015. http://www.epa.gov/region9/ag/ag-state.html.

"The State of the Birds 2011: Report on Public Lands and Waters." State of the Birds. Accessed February 2015. http://www.stateofthebirds.org/2011/State%20of%20the%20Birds%202011.pdf.

Suozzo, Andrea. "State Dairy Industry Dips Below 1,000 Farms." *Addison County Independent*, July 7, 2011. Accessed February 2015. http://www.addisonindependent.com/201107state-dairy-industry-dips-below-1000-farms.

"Tail Docking of Cattle Decried by Animal Advocates, Experts." *ABC News*, January 26, 2010. Accessed February 2015. http://abcnews.go.com/Blotter/watch-hidden-camera-video-tail-docking/story?id=9658808.

"Tylan Injection." Elanco. Accessed February 2015. http://www.elanco.us/products-services/beef/versatile-cattle-antibiotic.aspx.

"U.S. dairy farm numbers drop by 1,621— now total 51,481." Hoard's Dairyman. February 27, 2012. Accessed February 2015. http://www.hoards.com/blog_fewer-dairy-farms-left-business.

"The U.S. Dairy Industry: A Vital Contributor to Economic Development." Dairy Farming Today. Accessed February 2015. http://www.dairyfarmingtoday.org/SiteCollectionDocuments/economicfactsheet.pdf.

VanEgeren, Jessica. "Manure Digesters Seen as Solution to Runoff, but Drawbacks Persist." *LaCrosse Tribune*, April 30, 2014. Accessed February 2015. http://lacrossetribune.com/news/local/state-and-regional/manure-digesters-seen-as-solution-to-runoff-but-drawbacks-persists/article_2d517871-3b45-5f67-bcf7-b5e872935274.html.

Wallace, Richard L. "Bedding Choices: Mastitis Control and Cow Comfort." Accessed February 2015. http://livestocktrail.illinois.edu/uploads/dairynet/papers/2007%20dd%20Bedding%20Choices.pdf.

Chapter Fifteen: American Mountain Meadow

"A 2010 snapshot of U.S. dairying." Hoard's Dairyman. March 4, 2011. Accessed February 2015. http://www.hoards.com/?q=blog_U.S.%20dairying%20in%202010.

American Humane Association. Accessed February 2015. http://www.humaneheartland.org/.

Barnett, Lindsay. "California Legislature approves bid to ban tail-docking for cows." *Los Angeles Times—L.A. Unleashed*. September 3, 2009. Accessed February 2015. http://latimesblogs.latimes.com/unleashed/2009/09/california-state-legislature-approves-bid-to-ban-taildocking-for-cows.html.

"Bovine Growth Hormone (rBGH)/Recombinant Bovine Somatotropin (rBST)." Breast Cancer Fund. Accessed February 2015. http://www.breastcancerfund.org/clear-science/radiation-chemicals-and-breast-cancer/bovine-growth-hormone.html.

"CA Governor Approves Ban on Docking Cows' Tails." Animal Law Coalition. May 6, 2009. Accessed February 2015. https://animallawcoalition.com/ca-governor-approves-ban-on-docking-cows-tails/.

California Dairy Herd Improvement Association. Accessed February 2015. http://cdhia.org/.

"California Dairy Statistics 2011 Data." California Department of Food and Agriculture. Accessed February 2015. http://www.cdfa.ca.gov/dairy/pdf/Annual/2011/2012_Annual_2011_Data.pdf.

"California Firm Recalls Chicken Products Due to Possible Salmonella Heidelberg Contamination." United States Department of Agriculture—Food Safety and Inspection Service. Last modified September 30, 2014. http://www.fsis.usda.gov/wps/portal/fsis/topics/recalls-and-public-health-alerts/recall-case-archive/archive/2014/recall-044-2014-release.

"California has Four of the Top Five Dairy Cow Counties." *Farm Journal*, May 23, 2014. Accessed February 2015. http://www.agweb.com/farmjournal/article/california_has_four_of_the_top_five_dairy_cow_counties__NAA_Dairy_Today_Editors/.

"Changes in the Size and Location of U.S. Dairy Farms." United States Department of Agriculture, Economic Research Service. September 2007. Accessed February 2015. http://www.ers.usda.gov/media/430528/err47b_1_.pdf.

Charles, Dan. "How Foster Farms Is Solving The Case Of The Mystery Salmonella." *NPR*. August 28, 2014. Accessed February 2015. http://www.npr.org/blogs/thesalt/2014/08/28/342166299/how-foster-farms-is-solving-the-case-of-the-mystery-salmonella.

Crystal Creamery. Accessed February 2015. http://www.crystalcreamery.com.

"Dairy 2007, Part I: Reference of Dairy Cattle Health and Management Practices in the United States, 2007." United States Department of Agriculture. October 2007. Accessed February 2015. http://www.aphis.usda.gov/animal_health/nahms/dairy/downloads/dairy07/Dairy07_dr_PartI.pd.

Dohoo, I.R., L. DesCoteaux, K. Leslie, A. Fredeen, W. Shewfelt, A. Preston, and P. Dowling. "A meta-analysis review of the effects of recombinant bovine somatotropin." *The Canadian Journal of Veterinary Research* 67, no. 4 (2003): 252-264.

"Environment." Pew Commission on Industrial Farm Animal Production. Accessed February 2015. http://www.ncifap.org/issues/environment/.

"Environmental Stewardship." Foster Farms. Accessed February 2015. http://www.fosterfarms.com/about/stewardship.asp.

"Fact Sheet on Dairy Cows in Canada." Canadian Meat Council. November 2013. Accessed February 2015. http://www.cmc-cvc.com/sites/default/files/files/Fact%20Sheet%20Cull%20Dairy%20Cattle%20November%202013.pdf.

"Farm Labor: Background." United States Department of Agriculture, Economic Research Service. Last modified October 30, 2014. http://www.ers.usda.gov/topics/farm-economy/farm-labor/background.aspx#.VAa4hFa4lII.

Foster Farms. Accessed February 2015. http://www.fosterfarms.com/.

"Foster Farms Receives California's Highest Environmental Honor." Business Wire. Accessed February 2015. http://www.thefreelibrary.com/Foster+Farms+Receives+California's+Highest+Environmental+Honor+--+the...-a0139039522.

Haan, Mathew, Diana Stuart, and Becky Schewe. "Challenges and Benefits of Adopting Robotic Milking on Michigan Dairy Farms." Michigan Dairy Review. Accessed February 2015. https://www.msu.edu/user/mdr/vol17no3/challenges.html.

Holbrook, Stett. "Farming Communities Facing Crisis Over Nitrate Pollution, Study Says." Food & Environment Reporting Network. March 13, 2012. Accessed February 2015. http://thefern.org/2012/03/farming-communities-facing-crisis-over-nitrate-pollution-study-says/.

"The Humane Table." Foster Farms. Accessed February 2015. http://www.fosterfarms.com/about/humanetable.asp.

"International Competition: The Key to Success at Foster Farms Dairy." Alta Genetics. June 27, 2006. Accessed February 2015. http://www2.altagenetics.com/English/Whatsnew/20060825InternalCompetition.htm.

Marks, Robbin. "Cesspools of Shame." Natural Resources Defense Council and the Clean Water Network. July 2001. Accessed February 2015. http://www.nrdc.org/water/pollution/cesspools/cesspools.pdf.

Massey, Jim. "Taking robotics to another level: World's largest robotic rotary parlor being installed on southwest Wisconsin farm." *The Country Today,* August 25, 2014. Accessed February 2015. http://www.thecountrytoday.com/front_page/article_6c1f139a-2c5f-11e4-8a68-001a4bcf887a.html.

"Multistate Outbreak of Multidrug-Resistant *Salmonella* Heidelberg Infections Linked to Foster Farms Brand Chicken (Final Update)." Centers for Disease Control and Prevention. Last modified August 1, 2014. http://www.cdc.gov/salmonella/heidelberg-10-13/index.html.

National Dairy Herd Information Association. Accessed February 2015. http://www.dhia.org/.

"Number of Farms, Dairy Cows and Heifers." Government of Canada—Canadian Dairy Information Centre. Last modified September 19, 2014. http://www.dairyinfo.gc.ca/index_e.php?s1=dff-fcil&s2=farm-ferme&s3=nb.

"PA California Poultry Farm Pleads Guilty to Clean Water Act Violations." United States Environmental Protection Agency. January 23, 1998. Accessed February 2015. http://yosemite.epa.gov/opa/admpress.nsf/8b75cea4165024c685257359003f022e/1a96ac8b6affb5f985256595006fccc6!OpenDocument.

"rBGH." GRACE Communications Foundation. Accessed February 2015. http://www.sustainabletable.org/797/rbgh.

"rBGH (Posilac)—Breast Cancer and Prostate Cancer." Ethical Investing. Accessed February 2015. http://www.ethicalinvesting.com/monsanto/bgh.shtml.

"Real California Milk Facts." California Milk Advisory Board. Accessed February 2015. http://www.californiadairypressroom.com/Press_Kit/Dairy_Industry_Facts.

"Recombinant Bovine Growth Hormone." American Cancer Society. Last modified September 10, 2014. http://www.cancer.org/cancer/cancercauses/othercarcinogens/athome/recombinant-bovine-growth-hormone.

"Report on the Food and Drug Administration's Review of the Safety and Recombinant Bovine Somatotropin." U.S. Food and Drug Administration. Last modified July 28, 2014. http://www.fda.gov/AnimalVeterinary/SafetyHealth/ProductSafetyInformation/ucm130321.htm.

Rosson, Parr, Flynn Adcock, Dwi Susanto, and David Anderson. "The Economic Impacts of Immigration on U.S. Dairy Farms." National Milk Producers Federation. June 2009. Accessed February 2015. http://www.nmpf.org/files/file/NMPF%20Immigration%20Survey%20Web.pdf.

Rotz, C.A., C.U. Coiner, and K.J. Soder, "Automatic Milking Systems, Farm Size, and Milk Production." *Journal of Dairy Science* 86, no. 12 (2003): 4167-4177.

Sechen, Suzanne. "Bovine Somatotropin (bST)—Possible Increased Use of Antibiotics to Treat Mastitis in Cows." Food and Drug Administration, Center for Veterinary Medicine. October 30, 2013. Accessed February 2015. http://www.fda.gov/downloads/AnimalVeterinary/SafetyHealth/ProductSafetyInformation/UCM383073.pdf.

"The State of Farmworkers in California." Center for Farmworker Families. Accessed February 2015. http://www.farmworkerfamily.org/information/.

"Stewardship Profiles in California Agriculture." American Farmland Trust. Accessed February 2015. http://www.farmland.org/programs/campaign/documents/18-Fosterfarms.pdf.

Tucker, H. Allen. "Safety of Bovine Somatotropin (bST)." Extension. April 30, 2010. Accessed February 2015. http://www.extension.org/pages/11810/safety-of-bovine-somatotropin-bst.

"What It Means To Be American Humane Certified." *YouTube* video. Foster Farms. March 26, 2013. Accessed February 2015. https://www.youtube.com/watch?v=Ra0NRDjT9xQ.

Conclusion: The Future is Already Here

Abbott, Ellen. "New Farm Bill makes transition to organic dairy farming easier." *WVRO Public Media*, February 18, 2014. Accessed February 2015. http://wrvo.org/post/new-farm-bill-makes-transition-organic-dairy-farming-easier.

"Agricultural Disasters: Porcine Epidemic Diarrhea (PED)." Extension Disaster Education Network. Accessed February 2015. http://eden.lsu.edu/Topics/AgDisasters/Pages/PorcineEpidemicDiarrhea(PED).aspx.

"Avian Influenza A(H7H9) Virus." Centers for Disease Control and Prevention. Accessed February 2015. http://www.cdc.gov/flu/avianflu/h7n9-virus.htm.

"Avian Influenza (Bird Flu)." Centers for Disease Control and Prevention. Accessed February 2015. http://www.cdc.gov/flu/avian/gen-info/facts.htm.

Barclay, Eliza. "'Piglet Smoothie' Fed To Sows To Prevent Disease; Activists Outraged." *NPR*, February 20, 2014. Accessed February

2015. http://www.npr.org/blogs/thesalt/2014/02/20/280183550/piglet-smoothie-fed-to-sows-to-prevent-disease-activists-outraged.

Bittman, Mark. "Putting Meat Back in Its Place." *New York Times*, June 11, 2008. Accessed February 2015. http://www.nytimes.com/2008/06/11/dining/11mini.html.

"Chickens Do Not Receive Growth Hormones: So Why All the Confusion?" The Poultry Site. April 23, 2013. Accessed February 2015. http://www.thepoultrysite.com/articles/2812/chickens-do-not-receive-growth-hormones-so-why-all-the-confusion.

Covey, Stephen. *The Seven Habits of Highly Effective People*. New York: Free Press, 2004.

Cox, Billy. "USDA Establishes Naturally Raised Marketing Claim Standard." U.S. Department of Agriculture—Agricultural Marketing Service. January 16, 2009.

"Deregulated Foreign Meat Inspection Leaves U.S. Consumers Unprotected, USDA Records Reveal." Food & Water Watch. January 15, 2013. Accessed February 2015. http://www.foodandwaterwatch.org/pressreleases/deregulated-foreign-meat-inspection-leaves-u-s-consumers-unprotected-usda-records-reveal/.

"E.coli (Escherichia coli)." Centers for Disease Control and Prevention. Accessed February 2015. http://www.cdc.gov/ecoli/.

"Farmed Animals and the Law." Animal Legal Defense Fund. Accessed February 2015. http://aldf.org/resources/advocating-for-animals/farmed-animals-and-the-law/.

"Food: The EU and animal welfare: policy objectives." European Commission. Accessed February 2015. http://ec.europa.eu/food/animal/welfare/policy/index_en.htm.

"General Principles and Management Standards." Public Works and Government Services. Accessed February 2015. http://www.tpsgc-pwgsc.gc.ca/ongc-cgsb/programme-program/normes-standards/internet/bio-org/principes-principles-eng.html#a7.

Goetz, Gretchen. "Who Inspects What? A Food Safety Scramble." *Food Safety News*, December 16, 2010. Accessed February 2015. http://www.foodsafetynews.com/2010/12/who-inspects-what-a-food-safety-scramble/.

Harrison, Ruth. *Animal Machines*. London: Vincent Stuart Publishers, Ltd., 1964.

Hauter, Wenonah. "Citizen's Petition to Revoke Equivalency Determinations for the Canadian High Line Speed Inspection System; the Canadian HACCP-based Slaughter Inspection System; and the New Zealand Alternative Post-Mortem Meat Inspection Procedure." Petition, 2014. Accessed February 2015. http://documents.foodandwaterwatch.org/doc/HIMP_equivalency_petition_june_2014.pdf#_ga=1.186394665.1023616253.1402073225.

"Health and Food Safety: Animal Health and Welfare." European Commission. Accessed February 2015. http://ec.europa.eu/dgs/health_consumer/dyna/enews/enews.cfm?al_id=1332.

Hessler, Katherine, and Tanith Balaban. "Agricultural Animals and the Law," *GLSOLO* 26(5) (2009). Accessed February 2015. http://www.americanbar.org/newsletter/publications/gp_solo_magazine_home/gp_solo_magazine_index/agriculturalanimals.html.

Hirsch, Veronica. "Detailed Discussions of Legal Protections of the Domestic Chicken in the United States and Europe." Law Paper, Michigan State University College of Law, 2003. Accessed February 2015. https://www.animallaw.info/article/detailed-discussion-legal-protections-domestic-chicken-united-states-and-europe.

"Hormones and antibiotics in food production." EatRight Ontario. Accessed February 2015. http://www.eatrightontario.ca/en/Articles/Farming-Food-production/Hormones-and-antibiotics-in-food-production.aspx#.VJCz8ivF-So.
"IEC Report: 85 Per Cent of Eggs from Battery Cages." The Poultry Site. October 28, 2010. Accessed February 2015. http://www.thepoultrysite.com/poultrynews/21221/iec-report-85-per-cent-of-eggs-from-battery-cages.
King, Martin Luther, Jr. "Letter from a Birmingham Jail." 16 April 1963. Accessed February 2015. http://www.africa.upenn.edu/Articles_Gen/Letter_Birmingham.html.
Leonard, Annie. *The Story of Stuff.* New York: Free Press, 2010.
Livestock health care practice standard. *Code of Federal Regulations,* Title 7, subtitle B, chapter I, subchapter M, part 205, subpart C.
"Livestock in the Balance." The State of Food and Agriculture 2009. Accessed February 2015. http://www.fao.org/docrep/012/i0680e/i0680e02.pdf.
"Meatless Monday Goes Global!" Meatless Monday Global. Accessed February 2015. http://www.meatlessmonday.com/the-global-movement/.
Hatch, Steven. *The Contemplative John Muir.* Raleigh: Lulu, 2012.
"National Pork Board Statement of Porcine Epidemic Diarrhea Virus (PEDV)." National Pork Board. Accessed February 2015. http://www.pork.org/News/3904/NationalPorkBoardStatementOnPEDV.aspx.
Nestle, Marion. "USDA's new poultry inspection system, complicated but voluntary." *Food Politics,* August 4, 2014. Accessed February 2015. http://www.foodpolitics.com/2014/08/usdas-new-poultry-inspection-system-complicated-but-voluntary/.
Neuman, William. "As Supply Dwindles, Organic Milk Gets Popular." *New York Times,* December 29, 2011. Accessed February 2015. http://www.nytimes.com/2011/12/30/business/rising-production-costs-cause-organic-milk-shortage.html?pagewanted=all.
Orwell, George. *1984.* New York: Penguin, 1961.
Orwell, George. "Politics and the English Language." 1946. Accessed February 2015. https://www.mtholyoke.edu/acad/intrel/orwell46.htm.
Pollan, Michael. *The Omnivore's Dilemma.* New York: Penguin Books, 2006.
"Rising Number of Farm Animals Poses Environmental and Public Health Risks." Worldwatch Institute. Accessed February 2015. http://www.worldwatch.org/rising-number-farm-animals-poses-environmental-and-public-health-risks-0.
"Salmonella." Centers for Disease Control and Prevention. Accessed February 2015. http://www.cdc.gov/salmonella/.
Scully, Matthew. *Dominion.* New York: St. Martin's Press, 2002.
Sewell, Anna. *Black Beauty.* New York: F. M. Lupton, 1877.
Singer, Peter, and Jim Mason. *The Ethics of What We Eat.* Emmaus: Rodale, 2006.
Steinfeld, Henning, et al. *Livestock's long shadow: environmental issues and options.* Rome: Food and Agriculture Organization of the United Nations, 2006. http://www.europarl.europa.eu/climatechange/doc/FAO%20report%20executive%20summary.pdf.
Stevenson, Peter. "European Union Legislation on the Welfare of Farm Animals." Compassion in World Farming. January 2012. Accessed February 2015. http://www.ciwf.org.uk/includes/documents/cm_docs/2012/e/european_union_law_on_the_welfare_of_farm_animals_2012.pdf.
"Suit Accuses Kroger of Deceit on How Poultry Was Raised." *New York Times,* February 12, 2014. Accessed February 2015. http://www.nytimes.

com/2014/02/13/business/suit-accuses-kroger-of-deceit-on-how-poultry-was-raised.html?_r=0.

"Swine Flu (Swine Influenza A [H1N1 and H3N2v] Virus)." MedicineNet.com. Accessed February 2015. http://www.medicinenet.com/swine_flu/article.htm.

"Technical Note: Porcine Epidemic Diarrhea (PED)." United States Department of Agriculture. Accessed February 2015. http://www.aphis.usda.gov/animal_health/animal_dis_spec/swine/downloads/ped_tech_note.pdf.

Tischler, Joyce. "U.S. Lags Far Behind Europe in Protections for Farmed Animals." Animal Legal Defense Fund. August 15, 2011. Accessed February 2015. http://aldf.org/blog/u-s-lags-far-behind-europe-in-protections-for-farmed-animals/.

"USDA Petitioned to Block Imported Meat From Privatized Inspection Systems." *Food Safety News*, June 6, 2014. Accessed February 2015. http://www.foodsafetynews.com/2014/06/usda-petitioned-to-block-imported-meat-from-privatized-inspection-systems/.

"What is a CAFO?" U.S. Environmental Protection Agency. Accessed February 2015. http://www.epa.gov/region07/water/cafo/.